Data Science for Crime Analysis with Python

Andrew P. Wheeler

2024-06-23

Data Science for Crime Analysis with Python

ISBN 979-8-9903770-0-4 (ebook), ISBN 979-8-9903770-1-1 (paperback)

You can see my other work at https://crimede-coder.com/

Table of Contents

Preface

Python is an open-source, free, computer programming language. It allows you to write simple (or complex) programs, to allow your computer to accomplish tasks. For a brief example, here is a python code snippet that tells you the difference in the number of days between two dates. The lines that start with # are comments in python, the other lines perform different operations in python code. The text in the grey portion below is the python code, and the blue section text is the output of the program.

```
# importing library to calculate times
from datetime import datetime

# creating two datetime objects
begin = datetime(2022,1,16)
end = datetime(2023,1,16)

# calculating the difference
dif = end - begin

# printing the result
print(dif.days)
```

```
365
```

So this is a trivial program – you could figure out the number of days between the two dates using various tools. The power of being able to program in python is that you can write computer code to do (close to) any computation you want. A common example for a crime analyst may be querying a database and creating a table of crime counts year-to-date this year vs last. You can then run the code at will, and it will update the year-to-date stats on whatever frequency you want. Such a report in practice will just be chaining together short code examples like above into more complicated series of operations.

Who is this book for?

This book is aimed at individuals with no (or beginner) background in programming, but who are interested in using code to conduct quantitative analysis and automate tasks. The main intended audience of the book are crime analysts, but any individual looking to get started in coding and data analysis should find the content of the book useful. Besides crime analysts, those wishing to advance their career in a data science role or pursue graduate level research (who have a background in criminal justice), will find the book and its examples useful.

There are many current resources on using python on the internet – one can use a search engine to find various entirely free resources online. I commonly blog on technical computing at andrewpwheeler.com, which is free for anyone to read. These free resources are often haphazard though, and are very difficult for newcomers to understand and get started. Things like "How do I run a simple python script" or "How do I install a python library" are not typical topics covered in even introductory python materials online. This book is intended to make a singular resource for individuals in crime analysis to get started.

I intend this book to not only introduce code examples in python, but to also describe other necessary steps for beginners, such as setting up python environments and automating tasks using shell scripts. Don't worry if you do not understand what those are at the moment – they will be explained! I even spend time describing a typical project structure that is fairly standard across more professional software development. These are things not related to coding directly, but are necessary to be able to get started using python and use it effectively.

So this book fills a niche – an introduction to doing tasks in python of relevance to crime analysts. The book subsequently contains:

- installing and creating python environments
- an introduction to python programming
- working with tabular data using scientific libraries
- using SQL to query databases
- automating report creation and creating high quality tables and graphs

These are the necessary ingredients, both in terms of coding and doing more realistic projects, that allow one to become more productive in their regular tasks using python.

What this book is not

When approaching learning to write code and conduct data analysis, many books include *both* at the same time. This is often a mistake, as it can greatly increase the burden on those wishing to learn the material. This book *is not* meant as an introduction to crime analysis as a topic in general. For those who wish to learn basic statistics and analyses that crime analysts conduct, I would suggest to check out the course materials on my personal website, as well as the materials from the International Association of Crime Analysts (IACA). If demand is sufficient, I may create future books to cover intro statistics for crime analysts more thoroughly, so let me know if that is something you are interested in!

This book is aimed to get you started writing code and applying it to real tasks crime analysts need to conduct. I use realistic examples that a crime analyst may be interested in conducting, such as sending automated emails, making year-to-date tables, and creating line charts. But I do not discuss in detail things like the Poisson distribution for analyzing crime rates or why hotspot analyses is important.

While the material is definitely relevant to *everyone* who needs to conduct data analysis using python, I hope the use of more realistic crime analysis examples helps to better illustrate the utility of using python to conduct analyses in your day-to-day tasks as a crime analyst.

Why learn to code?

Crime analysts do much of their quantitative work in spreadsheets (e.g. Excel), and then a smaller number use additional tools, such as databases (e.g. Access, SQLServer), formatted documents (Word, Powerpoint, PDF), and GIS tools (such as ESRIs ArcMap). Why bother to learn python? I agree that Excel can be used to do amazing things with data, and many tasks are *exchangeable* between python and one (if not several) different tools.

The power of using programming, as opposed to tools that use a graphical user interface (e.g. point and click in the *GUI*), are:

- tasks can be fully automated
- tasks are fully documented

The first bullet point is an argument based on potential time savings for automating tasks. Say it takes you 30 minutes to do a task on a daily basis. If you take 100 hours to write python code to automate the task entirely, you have saved yourself time within 50 days from the automated process.

Many regular reports that crime analysts work on this time savings argument may not be compelling though. For example, when I worked as an analyst I had a monthly CompStat report with various graphs and maps. Using GUI tools it maybe took me 24 hours (three working days) to complete. Once I wrote code to automatically create the graphs, it was under a day task. But I maybe spent over 160 hours writing code to automate that task. It would take over a year to break even in terms of time savings.

Many regular reports crime analysts write will look like the latter; they will only be semi-regular, and so the time savings argument for automating via code is not as impressive. (Code automation makes more sense for time savings for things that need to be done more often.) Even in those cases of semi-regular reports though, I believe writing code to automate as much as possible is still worth it.

This is because of the second bullet point – tasks when written in code are by their nature fully documented. This provides the ability for an analyst to retrospectively say things like "this number looks weird, how did I calculate that?", or when a new analyst comes in and takes over the job, you can say "just go look at the scripts in folder X". Having standardized code provides a much more professional and transparent environment, which is helpful for you as an analyst as well as the organization as a whole.

It additionally allows you to scale your work. If you need to loan out your time forever for a particular task, even if only a single day per month for a particular report, you can only expand the scope of the work you do a certain amount. Being able to automate the boring stuff is a necessary step to free up your time to pursue other projects. It even allows you to go on vacation, and reporting requirements can still be fulfilled. Ultimately learning to code will likely make you more productive when conducting ad-hoc data analysis, as well as make you more marketable in a wider array of jobs (such as private sector data science jobs).

Why python?

The section above only describes why one would want to write code to automate tasks, it does not detail why to use python specifically (over say R or another statistical program). In addition to python, I have used SPSS (a program you need to pay for), and R (another open source statistical program) fairly extensively over my career. I have an R package, ptools, for regular functions of interest to crime analysts for example.

I have almost entirely migrated my personal coding to python, and do not use these other tools very often anymore. Again, python is very exchangeable with R for many tasks, but I prefer python at this point in my career due to its ability to manage entire projects, not just do a single task. In addition to this, many private sector data science positions focus almost entirely on python (and less so on R). So I believe in terms of professional development, especially if you have a goal for expanding your skills to pursue private sector data science positions, python is a better choice than R.

There are situations when paid for tools are appropriate as well. Statistical programs like SPSS and SAS do not store their entire dataset in memory, so can be very convenient for some large data tasks. ESRI's GIS (*Geographic Information System*) tools can be more convenient for specific mapping tasks (such as calculating network distances or geocoding) than many of the open source solutions. (And ESRI's tools you can automate by using python code as well, so it is not mutually exclusive.) But that being said, I can leverage python for nearly 100% of my day to day tasks. This is especially important for public sector crime analysts, as you may not have a budget to purchase closed source programs. Python is 100% free and open source.

How to read this book

I believe the optimal way to consume the material in this book is via a two step process. Your level of experience with python (either some or zero), will alter what materials you focus on and what ones you can likely skip. For everyone, I would suggest *skimming* each chapter briefly from the start, and understanding the high level goals each chapter is trying to teach.

This is how I personally consume technical material. You need to understand the high level goals that any particular piece of code is trying to accomplish before you can understand the finer technical details. If you cannot understand the high level goals, it will be very difficult to understand the technical details. It is also useful to understand what is possible – you do not need to point and click in excel to regenerate that CompStat report every month, you can write code to automate that (see Chapter 10).

The second part, after the skimming, is where it depends on whether you are neophyte to python, or whether you have some background experience in programming. For those neophytes with no experience, I would suggest you study in detail the entry chapters 1 through 4 in the book. A large problem in learning to run code is the "getting started running a simple example" problem – downloading a program and running commands is challenging for those who have never done it before.

This part – figuring out how to install python and run a simple command can be the most challenging hurdle to get started. Part of the challenge, as the author, is that everyone's systems is slightly different and changing over time. Instructions to get started tend to be idiosyncratic to your personal computer. Part of the reason I am writing this book is that most beginner materials do not even try to discuss this issue, and use tricks (such as using online platforms) to help people get started.

To accomplish real tasks crime analysts need for their jobs though, you cannot use the online platforms. Many individuals who are taught python in university courses use said online platforms (e.g. if you only have experience using Jupyter notebooks or only experience with Google Collab notebooks). You need to know how to download python and run it locally on your personal computer to be able to use it for work related tasks. But do not despair if you are having problems getting started! One technique professional software engineers use is called pair-programming – grab a friend who knows how to run python code (which can be me, or someone else in your network), look over their shoulder, and then have them look over your shoulder. This will help you get started in Chapter 1.

Chapters 2 through 4 introduce basic objects (strings, numbers, lists, dictionaries), and actions (conditional statements, looping, string substitution). These are very boring python basics – similar to how learning the normal distribution is boring in your introductory stats class, or learning algebra is boring in mathematics. They are the necessary building blocks though for understanding how to effectively write python code. Those with more entry level experience may feel comfortable skimming chapters 2-4; I would suggest to examine them in a cursory fashion at least though – there are likely a few things you did not know that will be introduced.

Chapter 5 is a section that even those with introductory experience often are not exposed. Writing your own functions and understanding how to import those functions are an important step from writing hobby code to creating a professional environment to develop work projects over time. Again, many individuals who have had a course in college on python programming are not exposed to this.

Chapters 6 through 9 are oriented around showing specific examples of working and presenting data that will be of wide interest, not only to those in crime analysis but those in any data oriented role. Chapter 6 shows the two main libraries to work with tabular data – `numpy` and `pandas`. Understanding the pandas library in particular is an important skill for those using python to conduct data analysis.

Chapter 7 shows how to use python to generate SQL queries. For those who are not familiar with SQL, *Structured Query Language*, SQL is used to pull data from an external database and into a pandas dataframe. This chapter I will also introduce different SQL statements, as in some scenarios it is better to do certain data analysis tasks in the database *before* you load the data into an in-memory pandas dataframe.

Chapter 8 introduces the `matplotlib` graphing library in python. Creating professional looking graphics is an important skill for data analysts. Generating high quality graphs is a signal to consumers of the quality of the work (for crime analysts these might be police officers, command staff, or the general public). Generating such graphs via code in python is a good way to control the look and consistency of graphs you produce.

Chapter 9 introduces Jupyter notebooks – notebooks offer a different environment than the terminal to run code. Jupyter notebooks can intermix plain text description, executable cells for code, and the output from those code executions (e.g. graphs and tables). This book, under the hood, is compiled from a series of Jupyter notebooks. I introduce Jupyter, as it is a convenient way to create standardized reports that contain different elements of data analysis.

The final chapter 10, *project organization*, discusses aspects of project management and workflow automation – the final necessary components to be able to take simple projects and really leverage python to help you do your job as a crime analyst. Now that you know how to write code, what does a project look like? There are standard ways you should organize your project, so when either you need to re-run the code, or others need to, they can understand the necessary components. This involves things like creating a README that has information to replicate the necessary environment to run the code, having functions documented and stored in a specific location, and a clear entry point that runs the code in an automated fashion.

The overall contents of the book are intended to go beyond "how to write python code", to giving individuals the end-to-end experience of creating realistic projects that can help crime analysts do their job. This involves more than just running a single script, but knowing how professionals do things like query a database, create reusable functions, and set up projects to automate different data analysis tasks over time.

These not writing code portions are what is severely lacking in current python programming how-tos, and is the main motivation to write the book.

My background

While getting my doctorate degree in criminal justice at SUNY Albany (between 2008 and 2015), I worked in several analyst roles. First, at the Division of Criminal Justice Services for New York state. That job mainly involved writing standardized reports based on the New York states criminal arrest history database. Then for several years I worked in-house at the Troy, NY police department as their lone crime analyst. Finally, I worked as a research analyst at the Finn Institute for Public Safety, a non-profit who collaborated in research projects with police departments in upstate New York.

I then was a professor of criminology for several years at the University of Texas at Dallas, from 2016 through 2019. During that time I wrote around 40 peer reviewed publications and collaborated on quantitative projects with police departments across the United States. I regularly presented this work at the IACA conference, and for a brief period was the head of the publications committee for IACA.

Currently (since late 2019), I have worked as a data scientist (in the private sector) for a healthcare company. My job now is to write software, focused on using predictive models in relation to healthcare claims data. While healthcare may seem quite different than crime analysis, many of the problems are fundamentally the same (working with health insurance claims is not all that different than working with crime reports). Examples of things I have built at my current private sector job are predictive models to identify when claims are overpaid, or when individuals are at high risk of a follow up heart attack.

The skills to build those predictive models are no different from work I have done forecasting crime in different areas, or identifying chronic offenders at high risk of committing future violence. So my personal experience as a crime analyst, a researcher, and then a private sector data scientist are what motivated me to write this book. I want to see some of the more advanced work in academia, as well as the software engineering practices in the private sector, trickle down more broadly into the crime analysis profession. An introductory book I believe is the best method to accomplish that goal.

Feedback on the book

For feedback on the contents of the book, you can send a message to me at https://crimede-coder.com/contact. Always feel free to send me feedback, suggest additional topics, or let me know of errors. If you are interested in more direct services, such as in person training for your analysts or direct consulting on projects you are working on, feel free to email me as well. Examples of past work I have conducted for various criminal justice agencies are program evaluation, redistricting, automating different processes, civil litigation consulting, and generating predictive models.

I have future plans to generate more advanced python content. These include books on:

- more advanced programming, data, and package management
- regression modeling
- machine learning applications in crime analysis
- GIS analysis with python

Feedback on content and letting me know what you are interested in helps me prioritize on this future work. And more sales of this individual book also give me motivation to write more – so tell your friends if you like it.

Thank you

Thanks are in order for several friends for reviewing early drafts of the book and providing critical feedback. Renee Mitchell gave me early feedback on drafts, and was one of the major impetuses to start this idea and finish it. Dae-Young Kim has been diligent in giving me very detailed feedback on each chapter, most importantly when I need to explain the code in more detail or my narrative is not consistent with the code examples. And I thank my wife for early feedback on the beginning chapters writing python code. (Easily the hardest part of coding is understanding how to run the code, not writing the code itself. A good test to see if your documentation is sufficient is to have someone who is *not* a programmer see if they can understand.)

Thank you all for the support, and for all of those who purchased pre-versions of the book.

This book is only possible based on various open source contributions. I am using the Quarto engine to render this book to different environments (PDF and EPUB), which itself uses LaTeX and Pandoc under the hood. Python itself is open source, and I extensively use tools by Anaconda. My thanks to all those individuals who help make the world go round behind the scenes.

1 Setting up python

First, before we can get started *writing code*, we need to set up python on your local machine. My suggestion for crime analysts is to install the Anaconda version of python, which can be downloaded for your machine at https://www.anaconda.com/download. I am writing this book on Windows, but most of my advice should also extend to Mac and Unix users (you can download Anaconda and run python on any of those operating systems). Where it might make a difference, I will provide a callout note. For example:

> Note
>
> When using different paths to file locations, Windows machines use backslashes, e.g. `C:\Users\andre`, whereas Mac and Unix machines use forward slashes, `/users/andre`.

Once Anaconda is installed, go ahead and open up the *Anaconda command prompt* (treat this just the same as any program you have installed on your machine, so on Windows can navigate the programs that pop up when clicking the Windows icon on the lower left). Once the Anaconda command prompt is open, it should look like something below.

Go ahead and type `python --version` into the command prompt, hit enter, and see what the results are. This will tell you if you have installed python correctly! My python version is `3.8.5`. Your version may be different than the one I wrote this book with, but the content I will be covering in this book it will not make a difference.

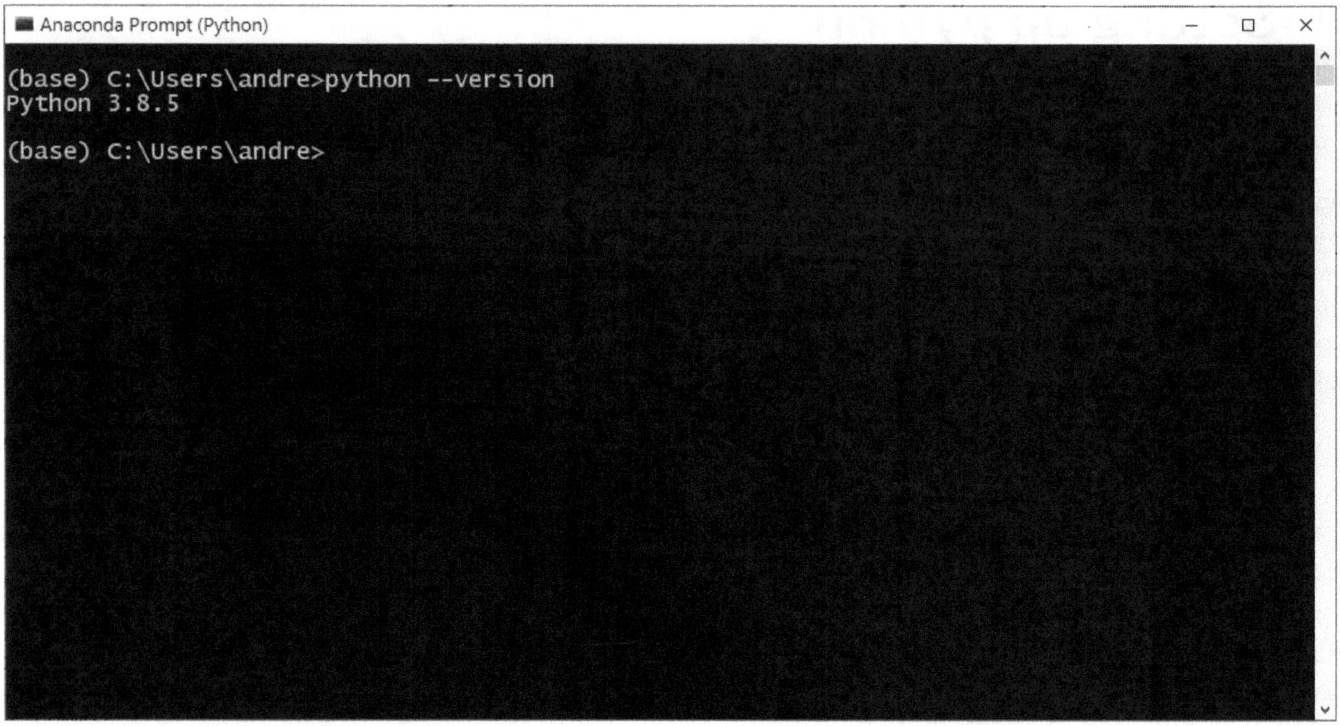

Anaconda Prompt (Python)

```
(base) C:\Users\andre>python --version
Python 3.8.5

(base) C:\Users\andre>
```

1.1 Running in the REPL

Now, we are going to run an interactive python session, sometimes people call this the *REPL*, read-eval-print-loop. Simply type `python` in the command prompt and hit enter. You will then be greeted with this screen, and you will be inside of a python session.

```
Anaconda Prompt (Python) - python
(base) C:\Users\andre>python
Python 3.8.5 (default, Sep  3 2020, 21:29:08) [MSC v.1916 64 bit (AMD64)] :: Anaconda, Inc
. on win32
Type "help", "copyright", "credits" or "license" for more information.
>>>
```

The cursor in the terminal should be located at the `>>>` part of the screen. Now at the prompt simply type `1 + 2`, hit enter, and it will output the answer:

```
Anaconda Prompt (Python) - python
(base) C:\Users\andre>python
Python 3.8.5 (default, Sep  3 2020, 21:29:08) [MSC v.1916 64 bit (AMD64)] :: Anaconda, Inc
. on win32
Type "help", "copyright", "credits" or "license" for more information.
>>> 1 + 2
3
>>>
```

Congrats, you have now written python code.

> Note
>
> The location of the cursor at the terminal at this point should be on the `>>>` line. You cannot go up to a prior line in the terminal, like you can in a word editor, but you can edit items on a single line in the terminal. So you can type `1 + 2`, and then before hitting enter hit backspace, and edit the line to be `1 + 3`.

1.2 Running a python script

Now lets make a simple python script, and call that script from the command line. First, navigate to any folder on your machine that you can add a file (see the note below for navigating to different locations via the command line). Here I navigated to the folder on my machine `B:\code_examples`, but your folder location will likely be different. Make a simple file, call it `hello.py`, in that same folder. You can initialize the file by simply writing `echo "" > hello.py` at the command prompt, or `touch hello.py` is easier if on a Mac/Unix machine. Or in the Windows OS make a text file and then rename the extension to `.py` instead of `.txt`.

> Note
>
> I will be giving advice for working at the command line in this book. It may seem complicated at first, but I only have a handful of commands memorized. It is important for project management to know exactly *where* you run commands from. Here are a few notes on using `cd` to navigate to different folders:
>
> - use `cd YourPath\YourFolder` to move to different folders, e.g. on Windows it might be `cd D:\Dropbox\Project`, whereas on Unix it might be `cd /Project/sub_folder`. On windows, to switch to a different drive, you can just type that drive letter (no `cd` at the front, e.g. if you just type `D:`, the command prompt will switch the directory to the `D:` drive.
> - If your path has a space in it, it needs to be quoted when using `cd`, for example `cd "C:\OneDrive\OneDrive - Uni\Folder"`. Without the quotes, the `cd` command will give an error.
> - use `cd ..\` (or `cd ../` on Unix/Mac) to move up a folder, e.g. if you are in `D:\Project\sub_project` and you type `cd ../` you will now be in `D:\Project`.
> - You do not need to type the full path to move down a single folder, so if you are in `D:\Project` and you type `cd .\sub_project` you will move down into the `D:\Project\sub_project` directory.
> - use the `pwd` command to print out the current directory.

In that `hello.py` file (which is just a text file), open it using whatever text editor you want. Then type these lines of code into that file, and save the file:

```
# This is a comment line
x = 'hello world'
print(x)

y = 3/2
print(y)
```

Now back at the command prompt, type in `python hello.py` and hit enter. You should see that it ran your python file, and printed out the results.

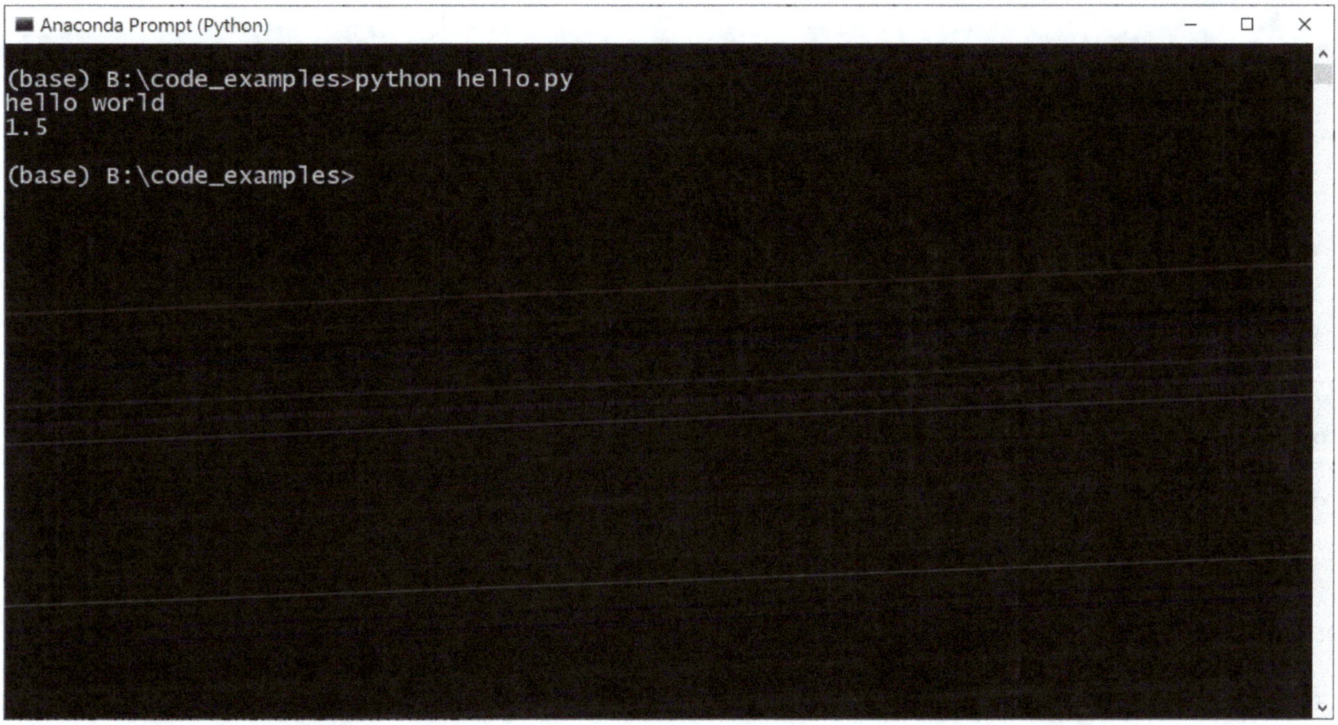

```
Anaconda Prompt (Python)

(base) B:\code_examples>python hello.py
hello world
1.5

(base) B:\code_examples>
```

Certain scenarios will dictate whether you are writing code in an interactive REPL session, or running code via scripts. Most of my work, I debug initial code using the REPL, and then save the finalized code and run the entire set of procedures through a script.

1.3 Some Extra Notes

I use Notepad++ to write much of my code. Here is what the prior set of code looks like, showing Notepad++ and the file itself:

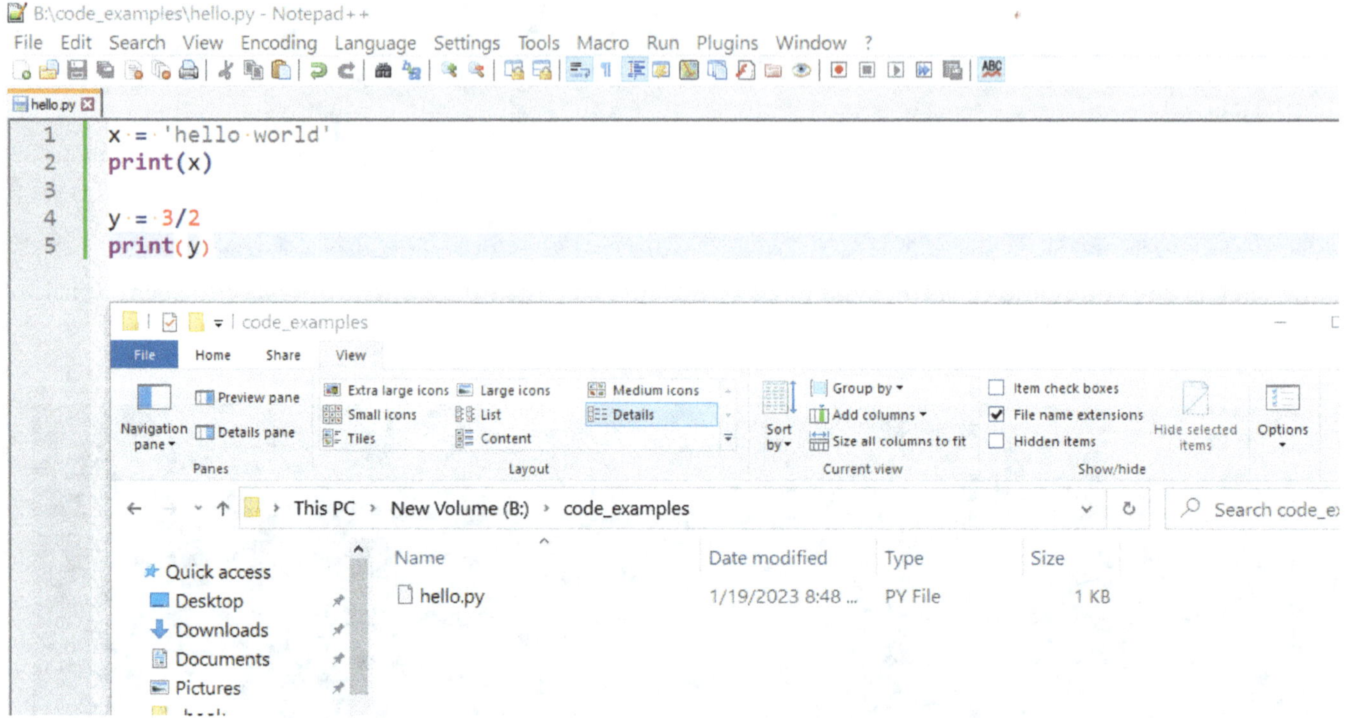

Notepad++ knows that a `.py` file is a python file, and so gives nice code formatting. It also allows you to set an option to see whitespace, which will become more important later when writing conditional code blocks in python. (You should not write code in a program that formats your text, such as Microsoft Word, as its auto-formatting can cause errors in your code.)

But there are other options to help you write python code. On occasion at my work I use Visual Studio (VS) Code, which is an entire IDE (integrated development environment). This just means it has extra stuff (like an embedded command prompt and github support). VS Code has extensions for python as well as several other languages. Another popular IDE for python is PyCharm. These again are mostly exchangeable, I like Notepad++ for its simplicity.

> Note
>
> Here are a few additional command line commands I use on a regular basis:
>
> - On Windows, to clear the terminal you can use `cls`, on Unix/MAC you can use `clear`
> - you can use `cat file.py` to print out the contents of a file to the terminal
> - you can use `mkdir` to make a new folder
> - you can pipe output to a file, e.g. `python hello.py > log.txt` would save the output of the command above to the `log.txt` file instead of printing to the terminal.
> - if you use `python script.py >> log.txt` this *appends* the output to log text file. Which is useful for repeated tasks that update over time.

Another option to write python code (especially for individuals who do scientific computing) is to use Jupyter notebooks. Many beginner python tutorials will suggest this. I will have a later chapter showing how to make a standardized report using Jupyter Notebooks, but I do not suggest this tool for beginners when first getting started. Many of the pieces of information I will discuss in this book about project management are difficult to control with Jupyter.

Now that you have the background to run python code, next chapter will cover more of the basics of how to write python code.

2 Getting started writing python code

While the prior chapter showed you how to run code, either in the REPL or via scripts, this chapter will get you started writing python code itself. The code examples shown via the text boxes I suggest typing into a REPL session to follow along, but you could also save them into `.py` scripts and run them directly as well. The grey portions are what you would type into the terminal, and the blue boxes are what should be printed out.

This chapter will walk through an introduction to working with numbers, strings, booleans, lists, and dictionaries. These *objects* are the building blocks of pretty much all code in python.

> Note
>
> On first read, this chapter (maybe all of them), will seem like a lot of information, and will likely be boring. I suggest to follow along in the REPL, but to *skim* the chapters fairly fast (especially Chapters 2, 3 & 4 on the basics of objects, strings, and looping). I have intentionally tried to be fairly comprehensive for the basics. When working on actual projects, you may need to revisit chapters to understand and re-acquaint yourself with the topics. You do not become an expert by reading a chapter one time, but by repeatedly writing code for your own projects over time.

2.1 Numeric Values

To get started, you can think about python code as *objects* and *operations* applied to those objects. So for example, if you run the python code:

```
x = 1
y = x + 1
print(y)
```

```
2
```

Here we first create an object, `x` that is *assigned* a value of 1 via the `=` symbol. We then create a second object, `y`, that is assigned the value `x + 1`. And then finally, we `print` the value of the `y` object. `print` is a function, whose only purpose is to output the value (or more specifically the string representation of that object), to the terminal (or whatever location you tell python to output its results to).

> Note
>
> In the REPL, it is also possible to just type a single item on a line and it will print the output to the terminal. So instead of `print(y)`, in a REPL session you could just type `y` and it will accomplish the same thing. In a script though, only `print(y)` would send the output to the terminal.

In the example above the objects were integer values. You can also have float numeric objects as well. Python, unlike some programming languages, does not force you to define the type of object before hand.

```
x = -1
y = 3.2
z = x/y
print(z)
```

```
-0.3125
```

When dealing with numeric values, python is smart and coerces `z` here to be a float value, even though it uses one integer as input (even with two integers, e.g. `z = 1/2`, in python `z` will be a float). You can see that here I did division, most of the math operations are similar to what you would type into any calculator, with the exception that powers use `**` instead of `^`:

```
# Showing off the different
# math operations
x = 2
print(x - 1)
print(x + 1)
print(x/2)
print(x*2)
print(x**3)       # x to the third power
print(x**(1/2))  # x to the 1/2 power (square root)
```

```
1
3
1.0
4
8
1.4142135623730951
```

One special operator in python is to modify an objects numeric value. So for example, to increment an objects value by one, you could do:

```
x = 1
x = x + 1
print(x)
```

```
2
```

But it is easier to use the special notation of `x += 1`:

```
x = 1
x += 1
print(x)
```

```
2
```

Note you can also do other mathematical operations, such as subtracting `x -= 1`, multiplication `x *= 2`, division `x /= 2`, or exponentiation `x **= 2`.

The final basic numeric examples to give are `//`, integer division, and `%`, the modulus operator. Integer division only gives the whole numbers in a division problem, and modulus is the remainder of the division problem.

```
x = 5
print(x//2)
print(x % 2)
```

```
2
1
```

Later chapters discussing libraries intended to work with tabular data (numpy and pandas), I will discuss numeric data processing in further detail. As you often don't want to do math on a single object, but a vector of multiple objects.

2.2 Strings

Another basic object you will often be using in python are strings. If you enclose a set of characters inside of quotes, that results in a string object. Here I even do addition of string objects, which concatenates the two strings together.

```
x = "ABC"
y = "de"
```

```
z = x + y
print(z)
```

```
ABCde
```

Strings can hold any alpha-numeric characters, so strings can hold text representations of numbers. Note that adding two strings is not the same as adding two numeric values! It concatenates the two strings together.

```
x = "1"
y = "3"
z = x + y
print(z)
```

```
13
```

You can coerce strings to numeric values via the `int` and `float` functions.

```
x = "1"
xi = int(x)
xf = float(x)
print(xi,",",xf) # can print multiple objects at once
```

```
1 , 1.0
```

And vice-versa you can coerce numeric values to be strings via the `str` function.

```
xi = 1
xf = 1.0 # if you use decimal, will be float

xsi = str(xi)
xsf = str(xf)

print(xsi + "|" + xsf) #concat the strings
```

```
1|1.0
```

> Note
>
> I have introduced three functions so far, `print`, `int`, `float`, and `str`. Functions in python take the form `function(input)`, so a particular name followed by two parentheses.

Strings can be enclosed either with single or double quotes. Note however that special characters in strings can cause issues. Backward slashes in windows path variables is one common example:

```
project_path = "C:\Project\SubFolder"
project_path # Note no print statement
```

```
'C:\\Project\\SubFolder'
```

> Note
>
> What gets printed to the console is not necessarily the same text representation of the object itself. For example, try `x = "Line1\nLine2"` and then type just `x` at the terminal and see what is printed out, vs typing `print(x)`. In this example, print interprets the line breaks in the string, whereas just typing `x` in a REPL session does not.

You can see that python inserted extra slashes! If you want the string exactly as you typed it, you can use the `r""` string option, which stands for *real* string.

```
project_path = r"C:\NoExtra\BackSlashes"
print(project_path)
```

```
C:\NoExtra\BackSlashes
```

If you want a multi-line string in python, you can use triple quotes. Note that this string has line breaks in the string object.

```
long_note = '''
This is a long
string note
over multiple lines
'''

print(long_note)
```

```
This is a long
string note
over multiple lines
```

If you have a very long string, such as a url, that you don't want to insert line breaks into, you can wrap the string in parenthesis. The python interpreter just turns it into a string in the end:

```
long_url = ("Pretend I am a super "
            "long url on a single line")

print(long_url)
```

```
Pretend I am a super long url on a single line
```

I have an entire chapter, Chapter 3, devoted to more advanced usage of strings.

2.3 Booleans

Boolean objects can only have two values, `True` or `False`.

```
print(1 == 1)
print(2 == 3)
```

```
True
False
```

You can see I use `==` to do an equality comparison. Remember a single `=` is for assignment. To use does not equal, the symbol is `!=`:

```
print('a' != 'a')
print('b' != 'c')
```

```
False
True
```

And then one can also use less than and greater than logic as well:

```
print(1 < 1)
print(1 <= 1)
print(2 > 1)
print(2 >= 1)
```

```
False
True
True
True
```

Note one can also do less than checks with strings, e.g. `'a' < 'b'` is `True`, but I don't use this very frequently. The example below shows a case that is probably not intended, as it accidentally compares string representations of numbers instead of numbers directly.

```
print('10' < '2')
```

```
True
```

Often you use a boolean to do conditional logic in python code. So you can have an `if` statement like below:

```
a = 1

if a == 1:
  print(a + 1)
```

```
2
```

A special part of python syntax is that white space is special. To denote that the `print` line is inside of the if statement, we append several spaces to the front. The number of spaces does not matter, it needs to be consistent though. (And you can technically also use tabs instead of spaces, but I highly recommend against that, as it can make editing the files a pain.)

Python uses `if`, `elif`, `else` for conditional statements. Here is an if and an else example:

```
a = 1

if a != 1:
  print('a does not equal 1')
  print(a + 1)
else:
  print('I am in the else part')
```

```
I am in the else part
```

The way these statements work, is that it checks if the first `if` statement is true. If that statement is true, it executes the code nested in the if part, *and then exits the code block*. If the `if` statement evaluates to False, it then executes the `else` statement block for all other cases.

Sometimes you want to chain together multiple checks, e.g. "if A, do Y, else if B, do X". To do that in python code, you would use `if` and then `elif`.

```
a = 1

if a != 1:
  print('a does not equal 1')
  print(a + 1)
elif a == 1:
  print('I am in the elif part')
else:
  print('I am in the else part')
```

```
I am in the elif part
```

If an `elif` is true, it executes that block and then exits, same as the `if` statement. So in the above code snippet, because the `elif` statement is true, it never gets to the `else` part of the conditional logic.

There is nothing that ties the different conditional statements together, so not all sections need to reference the same item:

```
a = 1
b = 'X'

if a != 1:
  print('a does not equal 1')
elif a == 2:
  print('a equals 2')
elif b == 'X':
  print('I am in the b check elif part')
else:
  print('I am in the else part')
```

```
I am in the b check elif part
```

And you can write subsequent further conditional logic inside of a conditional.

```
a = 1
b = 'X'
```

```
if a != 1:
  print('a does not equal 1')
  if b == 'X':
    print('b check in first if')
else:
  print('I am in the else part')
  if b == 'X':
    print('b check in elif')
```

```
I am in the else part
b check in elif
```

Sometimes in part of the condition, you want to do nothing. In those cases, you can use `pass` inside the condition.

```
a = 1

if a == 1:
  pass # This snippet will print nothing
else:
  print('I am in the else part')
```

You typically want the more common conditions first in a series of if statements, and less common conditions further down. So if the most common condition you do nothing, and only in rarer conditions you perform some action, this is a perfectly reasonable way to write your if statements.

These examples compare just two objects, but you can compose multiple conditional statements using `and` and `or`.

```
# and example
a = (1 == 1) and (2 == 2)
print(a)

b = (1 == 1) and (2 == 3)
print(b)

# or example
c = (1 == 1) or (2 == 3)
print(c)
```

```
True
False
True
```

There are short hand operators though, ampersand for `and` and the pipe for `or`:

```
# ampersand for and
a = (1 == 1) & (2 == 2)
print(a)

b = (1 == 1) & (2 == 3)
print(b)

# pipe for or
c = (1 == 1) | (2 == 3)
print(c)
```

```
True
False
True
```

You technically don't need the parentheses in the above examples, but I find it much easier to read the code that way and keep different terms together:

```
#       This is false          But this is true
a = ((1 == 2) & (2 == 2)) | (4 == 4)
print(a)
```

```
True
```

But most of the time, if possible, I just break it down in code and make the ultimate if statement line as simple as possible.

```
check1 = (1 == 2) & (2 == 2) # This is false
check2 = (4 == 4)            # This is true

if check1 & check2:
  print('Both check1 and check2 are true')
elif check1 | check2:
  print('At least one of check1 or check2 are true')
else:
  print('Neither check1 or check2 are true')
```

```
At least one of check1 or check2 are true
```

On occasion one does not use the math operators to generate the boolean statements, but `is` or `is not`. Perhaps the most common use of this is with the `None` object, what can be considered missing data in python.

```
x = None

if x is None:
  print('x has no value')
else:
  print('x has some value')
```

```
x has no value
```

Technically when you write `a is b`, it not only checks the objects values, but also that it references the *exact same object* in memory. Technically `x == None` will work above, but it is common practice to write it the `x is None` way. You can also check the opposite condition via `is not None`:

```
x = None

if x is not None:
  print('x has some value')
else:
  print('x is None')
```

```
x is None
```

For a final example, one can drop the comparison operators or if statements entirely. If you pass an "empty" object to an if statement, python checks to see if the object has any elements at all, and returns `True` if it does. So if you pass in an empty string, the if statement returns `False` here:

```
x = ''

if x:
  print('x has some value')
else:
  print('x is an empty string')
```

```
x is an empty string
```

This works for other python objects, such as empty lists and dictionaries, which will be illustrated in the subsequent section.

2.4 Lists

It is important for beginners to have an understanding of different objects that are containers for other objects. The first is a *list* object. A list contains multiple other objects, and you create a list by placing items inside brackets, and separating the items via commas. It is easier to show than to explain in words!

```
x = [1,2,3]
print(x)
```

```
[1, 2, 3]
```

Lists can hold different object types, it can contain both strings and numeric values in the same list for example. Here I also show that lists can contain other defined python objects.

```
a = 1
b = 'z'
x = [a,b,3]
print(x)
```

```
[1, 'z', 3]
```

You can split items in a list onto multiple lines in the python interpreter, it knows to look for the end bracket to know the input to the list is finished. So the resulting list below is exactly the same as `x = [1,2,3]`, just a different way to write it (it can be nice to split very long lists onto multiple lines for readability in your code).

```
# it is ok to define lists
# over multiple lines
x = [1,
     2,
     3]

# it is the same as earlier
# on a single line
print(x)
```

```
[1, 2, 3]
```

A common error when working with lists is to forget the comma. With numeric values this will often result in an error, but with strings it can sometimes not result in an error, since python will just implicitly concatenate the strings together. For an example:

```
# This is probably not what was intended
x = ['a' 'b', 'c']
print(x)
```

```
['ab', 'c']
```

So here the comma between the first two strings was omitted, so the resulting list is just two elements, with the first being `'ab'`.

> Note
>
> One benefit to writing lists on multiple lines, is you can use *column* editing in various text editors. In Notepad++, you can hold down the alt key and select multiple rows to edit at once. Most advanced text editors have a similar feature.

You can access individual items in a list via its index. Note in python that list indices start at 0, not at 1, so the first element of a list will be 0, the second element will be 1, etc.

```
y = ['a','b','c']
print(y[0])
print(y[1])
print(y[2])
```

```
a
b
c
```

You can also use *negative* indices to access items in reverse order in a list. So -1 accesses the last item in a list, -2 the second to last item, etc.

```
y = ['a','b','c']
print(y[-1])
print(y[-2])
```

```
c
b
```

Finally in terms of accessing multiple items in a list, you can use slice notation. So `x[1:3]` would access the 2nd and 3rd items in the list (it is equivalent to `[,)` notation is mathematics, so the left end of the slice is closed, and the right end of the slice is open.

```
y = ['a','b','c']
print(y[1:3]) # note it returns a list!
```

```
['b', 'c']
```

You can also use `x[1:]` to indicate, "give me the 2nd item in the list until the end", or `x[:3]` to "give me the first 3 items in the list.

One can create an empty list, simply by assigning `[]` to a value. Another useful trick to know is that you can do a boolean check for an empty list.

```
x = []

if x:
  print('The x list is not empty')
else:
  print('x is empty')
```

```
x is empty
```

An empty list has no objects, so if you try to access an object it will return an error. Above if you try to use `x[0]`, you will get an error `index out of range`.

Another boolean operation on lists is to check if an element is contained in that list.

```
x = ['a','b','c']

if 'd' in x:
  print('The list has a d element')
elif 'c' in x:
  print('The list has a c element')
else:
  print('x has neither d or c')
```

```
The list has a c element
```

One special piece of information you need to know about lists is that they are *mutable*. What does that mean exactly? It means that we can modify the contents of a list. So for example, we can replace a single item in a list.

```
y = ['a','b','c']
print(y)

y[1] = 'Z'
print(y)
```

```
['a', 'b', 'c']
['a', 'Z', 'c']
```

For a more complicated example, lists can point to other lists. Because lists are mutable, you can modify the inner list here, and the outer list reflects this change.

```
x = ['a','b','c']
y = [1, x]
print(y)

# if we alter x
# y points to
# the altered x list
x[1] = 'Z'
print(y)
```

```
[1, ['a', 'b', 'c']]
[1, ['a', 'Z', 'c']]
```

The way to think about this, the list `y` here does not actually contain the contents of `x`, it simply *points to* the `x` object. So if the `x` object gets changed, it points to that new `x` object.

This may seem quite in the weeds, but it is an important feature of the python programming language. It allows one to write many different algorithms in a simpler fashion when one can alter lists in place.

You can concatenate two lists together by adding them:

```
x = ['a','b','c']
y = [1,2]
z = x + y
print(z)
```

```
['a', 'b', 'c', 1, 2]
```

You can also make a repeated list via multiplication:

```
x = ['a','b','c']
y = [1]
print(y*3 + x*2)
```

```
[1, 1, 1, 'a', 'b', 'c', 'a', 'b', 'c']
```

Lists have several *methods* to modify their contents; two commons one used are sorting and reversing:

```
x = [3,1,2]

x.sort() # sorting the list
print(x)

x.reverse() # reverse ordering
print(x)
```

```
[1, 2, 3]
[3, 2, 1]
```

> Note
>
> When you sort or reverse a list, it does this operation and modifies the list *in-place*. So the code `y = x.sort()` is probably incorrect, as `x.sort()` returns nothing. So `y` does not equal the sorted list, but is `None`.

Methods are special functions that are tied to particular objects. So look like `object.method(input)`. They will always be demarcated from the base object by a period – so this means you cannot use a period in a variable name. For example if you type `q.i = 1` into the terminal it will return an error that `q is not defined`. These methods can take additional arguments (they are just like functions), but these examples just use the default. For example, you can sort in descending order:

```
x = [3,1,2]

x.sort(reverse=True) # passing arg
print(x)
```

```
[3, 2, 1]
```

Note

This is the first example I have shown for a function that has a *keyword* argument, so instead of `function(input)` it is `function(keyword=input)`. Functions can take multiple arguments, such as `function(input1,input2)`. In this scenario the order of the arguments matter, and you may use keyword arguments to distinguish between the inputs. I will go into more details on this in a subsequent chapter on defining your own functions.

You can also append or remove items from lists:

```
x = [3,1,2]

x.append('a') # appending an item to end of list
print(x)

x.remove(1) # removing an item
print(x)
```

```
[3, 1, 2, 'a']
[3, 2, 'a']
```

To find the specific location of an item in a list, you can use the index method:

```
x = ['a','b','c']

# will be 1, the 2nd item in a list
bindex = x.index('b')
print(bindex)
```

```
1
```

There are other methods to lists I have not shown here, if you run the `dir` command on an object, it will print out all of its potential methods. I encourage you to experiment yourself and see how the other methods, like `count` or `pop`, work.

```
x = ['a','b','c']

# you can look at the methods
# for a object using dir(object)
me = dir(x)

# there are many more! only
# printing a few to save space
```

```
print(me[-6:])
```

```
['index', 'insert', 'pop', 'remove', 'reverse', 'sort']
```

Sometimes you want a list like object, but you do not want it to be mutable (so you cannot do operations like change a single value, append, or sort the list). This may occur if you have a set of constants in your script, and you know they should never be altered. Placing them in a *tuple* is one way to ensure they don't accidentally get modified. Tuple's look mostly the same as a list, but use parentheses instead of square brackets:

```
y = (1,2,3)
y[1] = 5 # this will give an error
```

```
TypeError: 'tuple' object does not support item assignment
```

You can convert a list to a tuple via the `tuple` command:

```
y = [1,2,3]
z = tuple(y)
y[1] = 5 # this is ok
print(y) # can see y list is updated
z[1] = 5 # this is not, again cannot modify tuple
```

```
[1, 5, 3]
```

```
TypeError: 'tuple' object does not support item assignment
```

Or vice versa convert a tuple to a list via the `list` command:

```
z = (1,2,3)
y = list(z)
y[1] = 5 # this is ok
print(y)
z[1] = 5 # this is not, again cannot modify tuple
```

```
[1, 5, 3]
```

```
TypeError: 'tuple' object does not support item assignment
```

One last note about tuples, you can assign multiple objects at the same time. So you can do:

```
x, y = 1, 2
print(x,y)
```

```
1 2
```

This is called *tuple unpacking*. The reason it is called this is that when you *don't* unpack the multiple values separated by a comma, it returns a tuple.

```
t = 1, 2
print(t)
```

```
(1, 2)
```

And you can assign more than two values:

```
a,b,c,d = (1,'a',6,-1.2)
print(a,b,c,d)
```

```
1 a 6 -1.2
```

This can be convenient in various for loop examples (shown in Chapter 4), and when functions return multiple values (shown in Chapter 5). Besides this though, tuples are not as commonly used as lists in my experience. But one example use they have is illustrated in the next section, where one has a need to use immutable tuples for dictionaries.

2.5 Dictionaries

The second major container of items in python is a dictionary. So to access items in a list, it is just a set order, the first element is `mylist[0]`, the second element is `mylist[1]`, etc. Sometimes you want to be able to access the elements by simpler names, e.g. imagine you had a list to contain a persons information:

```
# using a list to hold data
x = ['Andy Wheeler','Data Scientist','2019']
```

To access the name item, you need to know it is in the 0 index, the title item is in the 1 index, etc. It is probably easier to refer to this data via a dictionary.

```
# using a list to hold data
d = {'name': 'Andy Wheeler',
     'title':'Data Scientist',
     'start_year': 2019}

print(d)
```

```
{'name': 'Andy Wheeler', 'title': 'Data Scientist', 'start_year':
2019}
```

Now, it is easier to access an individual item via `dict[key]`, so if I only want the name, I just reference that explicitly:

```
# Grabbing the specific name element
print(d['name'])
```

```
Andy Wheeler
```

The terminology for dictionaries is that they have *keys* that reference *values*. Values can be anything: numeric values, strings, lists, other dictionaries, etc. Keys however need to be *immutable*, even though dictionaries themselves are mutable (so you cannot use a list as a key, but you can use a tuple). Here we can modify the values of the original `d` dictionary I created. You can also add in a new element once the dictionary is created.

```
# Modifying the start_year element
d['start_year'] = 2020

# Adding in a new element tenure
d['tenure'] = 3

print(d['name'])
print(d['title'])
print(d['start_year'])
print(d['tenure'])
```

```
Andy Wheeler
Data Scientist
2020
3
```

Most often people use strings for keys, since the main benefit of dictionaries over lists is to have a name for referencing specific objects. But it can be a number as well. So say you had numeric ids in another database referencing specific locations, it may make sense to write your dictionary using those same numeric identifiers.

```
# You can have numeric values
# as a dictionary key
d = {} # can init a dict as empty
d[101] = {'address': 'Penny Lane', 'tot_crimes': 1}
d[202] = {'address': 'Outer Space', 'tot_crimes': 42}

# These show a dictionary inside of a dictionary
print(d[101])
print(d[202])
```

```
{'address': 'Penny Lane', 'tot_crimes': 1}
{'address': 'Outer Space', 'tot_crimes': 42}
```

And similar to empty lists, an empty dictionary will return `False` in a boolean if statement:

```
d = {} # can init a dict as empty

if d:
  print('The dictionary d is not empty')
else:
  print('The dictionary d is empty')
```

```
The dictionary d is empty
```

There are *many* types of more complicated objects in python, the very first example in the preface showed an example working with *datetime* objects. But under the hood, they are often just a container to conduct different operations on the objects I listed above: numeric values, strings, lists, and dictionaries.

3 Working with strings

In Chapter 2, I introduced string objects. Strings come up in many scenarios, regular examples I have encountered are creating standardized reports, generating SQL to query databases, and manipulating values in a field to extract more information. This chapter will walk you through first string manipulation, second searching strings, and the third section will give different methods to generate strings.

3.1 Manipulating strings

Elements of a string can be accessed similarly like elements of a list, to slice off portions you can access their indices:

```
x = 'This is a string'
print(x[0])
print(x[1])
print(x[-1])
print(x[2:5])
```

```
T
h
g
is
```

So if you want the first three elements of a string, you would do `x[:3]`, and for the final three elements of a string you can use `x[-3:]`.

Typically if you want the end of the string, you do not want to include potential whitespace. So if you have a string `'abcd   '`, and you access the final three elements of that string, you will just get three spaces. Strings have a method, `.strip()`, to eliminate whitespace.

```
x = '  abc '
print(x.strip())
```

```
abc
```

Note that this eliminates the spaces in both the beginning and the end of the string. There are methods for just stripping the elements on the left or right of the string, `.lstrip()` or `.rstrip()` respectively. You can also pass in an argument to `.strip()` to specify the particular characters to strip (default it strips spaces and line ending characters):

```
x = '  abc__'
print(x.strip("_"))
```

```
  abc
```

So here the spaces at the front of the string stay, but the underscores at the end of the string are eliminated. Since `x.strip()` itself returns a string, you can chain together different methods. So if you wanted the final three characters eliminating whitespace:

```
x = '  abcdef  '
print(x.strip()[-3:])
```

```
def
```

You can also chain other methods, for example you can switch the case of strings to uppercase or lowercase via `x.upper()` or `x.lower()`.

```
x = '  abcdef  '
print(x.upper().strip())
```

```
ABCDEF
```

You can check out all of the methods for strings by typing in `dir(str)`, another method instead of setting the strings entire case to lower or upper letters is to set the string to title case:

```
x = 'andrew wheeler'
print(x.title())
```

```
Andrew Wheeler
```

3.2 Searching strings

Sometimes we want to search (and often extract), different elements of strings. To find a specific substring within a string, you can use the `.find()` method to return the start index location:

```
x = 'abcde'
print(x.find('bc'))
```

```
1
```

Note that find returns -1 when the substring is not found:

```
x = 'abcde'
print(x.find('z'))
```

```
-1
```

Imagine you have a field in a database that has zipcodes. For geocoding, you just want the zip5, but some of the fields have zip5+4. There are different ways you may handle the situation. One is to search for the delimiter, and only return the elements *before* the +4:

```
zip = '75203-1234'

dash = zip.find("-")
zip5 = zip[:dash]
print(zip5)
```

```
75203
```

But note what will happen if you supply a zip code without the dash:

```
zip = '75203' # no dash

dash = zip.find("-")
zip5 = zip[:dash]
print(zip5)
```

```
7520
```

This does not generate an error message, but is obviously not what you intended! Since `zip.find("-")` does not find the dash, it returns `-1`. So the string slicing is then `zip[:-1]`, which returns the entire string *minus* the final character.

You may solve this via an if statement:

```
zip = '75203' # no dash

dash = zip.find("-")

if dash > -1:
  zip5 = zip[:dash]
else:
  zip5 = zip

print(zip5)
```

```
75203
```

But here if your zipcodes are so consistently formatted like this, it will be easier to just grab the initial 5 characters to begin with:

```
zip = '75203-1234'

zip5 = zip[:5]
print(zip5)
```

```
75203
```

And that will work for both for just zip5's as well as zip+4s formatting versions. Another way of grabbing out begin/end parts of delimited strings, is to the use `.split()` method. This will split a string into a list of different parts.

```
zip = '75203-1234'

print(zip.split("-"))
```

```
['75203', '1234']
```

Note that even with no dashes in the zipcode string, you will still get a list, just with a single element.

```
zip = '75203'

print(zip.split("-"))
```

```
['75203']
```

So here if you wanted zip5, just using `zip.split("-")[0]` would work, for both zip+4s as well as just zip5s. You can split a string on whatever delimiter you want, default it splits on whitespace.

```
name = 'Andrew Wheeler'

print(name.split())
```

```
['Andrew', 'Wheeler']
```

So if you had a mix of names with middle names, you could grab the first element of the list, as well as the last element of the list:

```
name = 'Andrew P Wheeler'

name_split = name.split()
first_name = name_split[0]
last_name = name_split[-1]

print(first_name)
print(last_name)
```

```
Andrew
Wheeler
```

And this will work even if you pass it a name with no middle name/initial, e.g. it works for `name = 'Andrew Wheeler'` as well as `name = 'Andrew P Wheeler'`. Unfortunately on occasion in name databases in my experience, people have additional whitespace and extra characters, e.g. "Guy La Salle", where "La Salle" is the last name. So this example would fail the parsing above.

Sometimes when you are searching for elements in a string, you just want to eliminate them. Say you had names with apostrophes in them, and you wanted to strip them out. You can use the `.replace()` method:

```
name = "Andrew P W'eeler"

print(name.replace("'",""))
```

```
Andrew P Weeler
```

A final example of searching strings, for more complicated search patterns, you can use *regular expressions*, more commonly known as regex's. Python has a standard library for regex's. Here I illustrate searching a string for both "Andy" and "Andrew". This is similar to the find example earlier, but you can use an or condition, via the pipe, and it returns the specific match.

```
import re # the standard regex library

s1 = "Andy Wheeler was xxxx"
s2 = "yyyy happened Andrew Wheeler"
s3 = "no andy"


print(re.findall("Andy|Andrew",s1))
print(re.findall("Andy|Andrew",s2))
print(re.findall("(Andy|Andrew)",s3))
```

```
['Andy']
['Andrew']
[]
```

The power of regex's is that you can capture different groups in the results, and search for more complicated patterns. Here is an example that searches a string for a license plate of the form 3 alpha characters, a dash, and then four digits. And returns each of the two alpha characters and groups as separate components.

```
import re

l1 = "xxxxx ABC-1234 yyy"
l2 = "ABC-1234"
l3 = "no license plate"

# Can save the regex for multiple use
re_lp = re.compile("([a-zA-Z]{3})-([0-9]{4})")

print(re.findall(re_lp,l1))
print(re.findall(re_lp,l2))
print(re.findall(re_lp,l3))
```

```
[('ABC', '1234')]
[('ABC', '1234')]
[]
```

So if your head spins looking at `([a-zA-Z]{3})-([0-9]{4})`, no worries, we all have a hard time looking at the end result of a complicated regex and trying to parse it. Starting from the beginning, if you want to search for 3 alpha characters in a row, you can use `[a-zA-Z]{3}`, the items in the bracket are shorthand for looking at all alpha characters in both upper and lower case. The 3 in brackets is saying to search for three elements. We can do something similar for numbers, which is `[0-9]{3}`. Putting it all together, we enclose different portions in parentheses to capture those groups, and look for a dash in-between.

> Note
>
> In these examples, I use `import re` at the beginning of each code block. In a real python script or REPL session, you only need to import the library one time. I do this here just to make each code snippet stand alone and runnable.

I have one additional regex that I commonly use. When I want to eliminate any non-alpha characters, such as apostrophe's in names. You can use a regex to look for anything *that is not* an alpha character (or space). You do this via `[^a-ZA-Z\s]` – the carrot at the beginning of the bracket signifies any character that does not meet the search pattern. The `\s` is a special signifier in regex that says look for a space. Then you can use `re.sub`, similar to how we used `str.replace` earlier:

```
import re

n1 = "Andy Wheeler"
n2 = "Andy, Wheeler"
n3 = "Andy W'eeler"

# use real string when using slashes
search = r'[^a-zA-Z\s]'
replace = '' # strip out entirely

print(re.sub(search,replace,n1))
print(re.sub(search,replace,n2))
print(re.sub(search,replace,n3))
```

```
Andy Wheeler
Andy Wheeler
Andy Weeler
```

> Note
>
> In the PDF version of the book, the carrot symbol, "^", sometimes does not copy and paste properly. So if this example is giving you troubles, make sure the carrot is correctly copied into your terminal or code editor!

Regex's are a difficult subject (and would require an additional chapter on their own to go into full detail). I more often use the split method above, to work on tinier substrings, and then on occasion use regex's. But often when working with the tinier strings, the simpler string methods of find and replace work quite well.

3.3 Generating strings

This section shows techniques for generating new strings. Why might you want to do this? Common examples in my work are inserting text into tables/reports, or to print out certain messages to the log when running scripts. Another is creating parameterized SQL strings to query databases. I will show examples of each.

First, is using what are called f-strings to do text insertion:

```
x = 'Andy Wheeler'
res = f'Writers name is {x}'
print(res)
```

```
Writers name is Andy Wheeler
```

So here we have our f string template for a string `f'Writers name is {x}'`. The part in the brackets, `{x}`, gets expanded to insert the actual value of x. So say you had a report title, `Year-to-date {Month}-{Day}-{Year}`, you can dynamically insert the current month/day/year. Here I illustrate using the datetime library to do this (this output will be different for you, as it will be tied to the current day you run the code!):

```
from datetime import datetime

# Get current date
current = datetime.today()

# Extract day/month/year
day = current.day
month = current.month
year = current.year

# Put them in string
title = f'Year-to-Date Stats as of {month}-{day}-{year}'
print(title)
```

```
Year-to-Date Stats as of 2-21-2024
```

So you can see in this example, even though the variables `day`, `month`, and `year` are numeric values, they still get inserted into the f-string. When you have float values though, they can sometimes insert very superfluous decimal values:

```
x = 1/3
res = f'Too many decimals {x}'
print(res)
```

```
Too many decimals 0.3333333333333333
```

One solution is to round the value, e.g. instead of `{x}` use `{round(x,2)}` to round to `0.33`. But an even simpler solution is to use string formatting codes. Here you can specify to format the float value to only display two decimals via `{x:.2f}`:

```
x = 1/3
res = f'Only two decimals {x:.2f}'
print(res)
```

```
Only two decimals 0.33
```

For long numeric values, such as large dollar values, you can print out comma delimited thousand markers as well. Here I also round to the whole integer value, not showing any decimal places:

```
x = 1025.6578910
res = f'Value ${x:,.0f}'
print(res)
```

```
Value $1,026
```

With f-strings, you can do the same operations as all normal strings. You can split them on multiple lines with triple quotes:

```
title = 'Data Science for Crime Analysis with Python'
author = 'Andrew Wheeler'
res = f'''Title: {title}
Author: {author}'''

print(res)
```

```
Title: Data Science for Crime Analysis with Python
Author: Andrew Wheeler
```

Or you can use real quote strings as well:

```
subject = 'Date Science'
tool = 'Python'

# Without real string, \n will be interpreted
# as a line break
title = f'{subject} for Crime Analysis\nwith {tool}'
print(title)

# With fr, will treat that as a real string
# and not interpret \n as a line break
title2 = fr'{subject} for Crime Analysis\nwith {tool}'
print(title2)
```

```
Date Science for Crime Analysis
with Python
Date Science for Crime Analysis\nwith Python
```

In a prior section, I presented how if statements need to have white space to identify the scope of an if-else blocks. Sometimes with very long, multi-line strings, this can look strange, as it breaks up the visual indenting (although still works).

```
x = 1

if x > 0:
  new_val = x*1000
  long_str = f'''I am a long
multi-line string
The value is {new_val:,}'''
  print(long_str)
elif x <= 0:
  new_val = 0
  long_str = f'''I am a long
multi-line string
The value is {new_val}'''
  print(long_str)
```

```
I am a long
multi-line string
The value is 1,000
```

The generated string in this example needs an element, `new_val`, which is only calculated conditional inside of the if-elif statement. So you cannot generate the string outside of the boolean functions. An option here is to make a string template, and then insert that value inside of the boolean function. There is a special library for this in python:

```
from string import Template
x = 1

long_temp = Template('''I am a long
multi-line string
The value is $val''')

if x > 0:
  new_val = x*1000
  long_str = long_temp.substitute(val=new_val)
  print(long_str)
elif x <= 0:
  new_val = 0
  long_str = long_temp.substitute(val=new_val)
  print(long_str)
```

```
I am a long
multi-line string
The value is 1000
```

I commonly use this to generate SQL statements where I want to parameterize different elements, such as the type of crime queried and the date range of queries. Below is an example.

```
from string import Template

sql = Template("""/* Example query */
SELECT
  *
FROM table
  WHERE crimetype = '$crimetype'
  AND date BETWEEN '$begin' AND '$end'
""")

query = sql.substitute(crimetype='Robbery',
                       begin='2019-01-01',
                       end='2019-12-31')
print(query)
```

```
/* Example query */
SELECT
  *
FROM table
  WHERE crimetype = 'Robbery'
  AND date BETWEEN '2019-01-01' AND '2019-12-31'
```

Chapter 7 will discuss more examples of using python + SQL to generate queries from databases.

For these examples, you could also generate a simple f-string template. This again is often useful if you need to expand the string repeatedly for different elements.

```
# Note this is not an f-string, but has the brackets
message = "Hello {placeholder}, here is your custom message"

# list of individuals
names = ['Bart','Lisa','Homer']

# looping over and creating new message
for name in names:
    res = f"{message}".format(placeholder=name)
    print(res)
```

```
Hello Bart, here is your custom message
Hello Lisa, here is your custom message
Hello Homer, here is your custom message
```

These examples there is not much difference between the string Template and f-string placeholders. One scenario (which does not occur often), is that if you have missing data, you may wish to use the string template `.safe_substitute` method in-place of `substitute`. Personally I like using string Templates for code clarity for longer strings, and f-strings for shorter strings.

Instead of inserting text into a template, sometimes you have multiple strings you want to add together to make a larger string in the end. I showed this previously when first introducing strings, how you can concatenate strings by adding them together.

```
street = '100 Main St.'
city = 'Raleigh'
state = 'NC'

full_add = street + "," + city + "," + state
print(full_add)
```

```
100 Main St.,Raleigh,NC
```

If you need to do this for a large set of strings, it may be easier to use the `.join()` method. This method is a little bit different though, you do the operation on the delimiter you want (here a comma), and then pass the list as an argument to the method.

```
street = '100 Main St.'
city = 'Raleigh'
state = 'NC'

add_list = [street,city,state]
res = ",".join(add_list)
print(res)
```

```
100 Main St.,Raleigh,NC
```

If you want to stack together multiple string elements into a single field, sometimes I find it useful to use as a delimiter a pipe symbol, |. As the pipe symbol is uncommon in most strings, you can split the elements back later if you need to.

```
c1 = "Witness A comment"
c2 = "Victim comment"
c3 = "Witness B comment"

comment_list = [c1,c2,c3]
res = "|".join(comment_list)
print(res)
```

```
Witness A comment|Victim comment|Witness B comment
```

And note that depending on the example, you may wish to have spaces separating the elements, e.g. use `" | ".join(comment_list)` instead of `"|".join(comment_list)`.

Sometimes you want to generate strings with leading characters, such as 0's. You can use the `.zfill()` method to do this.

```
val1 = '1'
val2 = '2'
val100 = '100'

print(val1.zfill(3))
print(val2.zfill(3))
print(val100.zfill(3))
```

```
001
002
100
```

4 Iterating over objects

This chapter introduces *iteration*. This mostly includes looping over objects. Do you need to send an email to a list of students you have in your class? A typical way to do that would be to have your students email addresses in a list, and then use a *for-loop* to iterate over that list and send an email one at a time. Or do you need to create a series of charts for different crime types – one way to do that would be to have a list of different crime types, and loop over them to generate the repeated charts.

4.1 Loops

There are two common types of loops in python, for loops and while loops. For loops iterate over a specific number of objects.

```
# have a list of email addresses
email = ['p1@gmail.com','p2@hotmail.com','p3@apple.com']

# looping over those strings in the list
# and printing them out
for e in email:
  print(e)
```

```
p1@gmail.com
p2@hotmail.com
p3@apple.com
```

Notice that *whitespace* is important here. The command below the loop, `print(e)`, has two spaces before the print statement. The number of spaces does not matter, it just needs to be consistent (it is the same indenting logic when using conditional statements as well).

While loops iterate until a condition is met, which can be a variable number of times until the terminating condition is met:

```
v = 10 # init an object to a value of 10

# iterate as long as the objects value is less than 17
while v < 17:
```

```
    v += 2
    print(v)
```

```
12
14
16
18
```

I do not come across while loop scenarios very often in my work – they occur in some statistical scenarios (iterating until convergence in an algorithm), or repeatedly trying some task until it works at least one time (such as querying a web API). So I focus the rest of the chapter on for loops. Even with those examples, a for loop with a fixed number of iterations (to prevent getting stuck in a while loop forever) and using a `break` statement is a common pattern:

```
v = 10 # init an object to a value of 10

# fixed number of iterations, but breaks
# if condition is met, max 100 iterations
for _ in range(100):
  v += 2
  print(v)
  if v >= 17:
    break
```

```
12
14
16
18
```

In for loops, using the underscore, `_`, is a common convention if you don't want to actually use the resulting value for any operation. Many python introductions start with iterating over a series of numbers, `range(5)` will loop over the integers 0 to 4 (remember, python starts indices at 0, not at 1):

```
# looping over numbers 0-4
for i in range(5):
  print(i)
```

```
0
1
2
3
4
```

> Note
>
> Technically `range(5)` does not produce a list, but *an iterable* object. (I know, I am using *iterating* a lot – apologies!) The difference between `[0,1,2,3,4]` vs `range(5)` is that the latter does not create the object in memory all at once. It iterates over the list of numbers, and *yields* each sequential number one at a time. For small lists this does not matter, but if you loop over `range(1000000)` it is more memory efficient to not hold the list of 1 million numbers in your computer memory all at once.

Sometimes people have a set of code that looks like this, calculating the length of a list using `len`, then using `range` to iterate over the numbers, and then extracting out each element of the list:

```
alpha = ['a','b','c']
len_x = len(alpha)

# looping over numbers
for i in range(len_x):
  print((i,alpha[i]))
```

```
(0, 'a')
(1, 'b')
(2, 'c')
```

But there is a simpler way to do this, it is using the `enumerate` function:

```
alpha = ['a','b','c']

# generating values/numbers at same time
for i,a in enumerate(alpha):
  print((i,a))
```

```
(0, 'a')
(1, 'b')
(2, 'c')
```

This makes the code simpler, in that I am just working with the elements directly, `a`, instead of having to do the list slicing, `alpha[i]`. Since `enumerate` returns both the index value and the actual item, this for loop uses tuple unpacking to reference each element individually in the for loop.

Even though it is not needed here in these simple of examples, another approach to incrementing a counter within a loop is to use the `+= 1` notation I described in Chapter 2:

```
alpha = ['a','b','c']

# calculating counter
c = 0 # starts at 0 so will be 1/2/3
for a in alpha:
  c += 1
  print((c,a))
```

```
(1, 'a')
(2, 'b')
(3, 'c')
```

You can loop over several lists at the same time using the `zip` function:

```
lx = [1,2,3]
ly = [4,5,6]

# iterating over two lists
for x,y in zip(lx,ly):
  print((x,y,x+y))
```

```
(1, 4, 5)
(2, 5, 7)
(3, 6, 9)
```

And similar to enumerate, this uses tuple unpacking to give different names of x,y to the two items. If you had three lists you should unpack three names e.g. `for x,y,z in zip(l1,l2,l3)` etc.

> Note
>
> You can see I commonly use short names in for loops, such as `for s in some_list`. This is just a convention to make code easier to read. You can name the for object anything you want that is a valid object name in python (so no spaces or certain special characters). Shorter names are often easier to work with though, and still tend to be relatively clear what they refer to in the code. If you used something like `for some_long_name in some_list:` you would have to then write code to work with `some_long_name` instead of `s` in the subsequent code.

You do not *need* to unpack the resulting zipped objected though, it is just implicitly cast to a tuple:

```
lx = [1,2,3]
ly = [4,5,6]
```

```
# zip returns an iterator of tuples
for combo in zip(lx,ly):
  print(combo)
  print(combo[0]+combo[1])
```

```
(1, 4)
5
(2, 5)
7
(3, 6)
9
```

Note one gotcha here – if the lists are of different lengths, zip stops after the shorter list:

```
lx = [1,2,3]
ly = ['a','b']

# this stops after shorter ly has ended
# it does not print 3
for x,y in zip(lx,ly):
  print(x,y)
```

```
1 a
2 b
```

You can loop over other objects as well. Tuples are essentially the same as a list (just again you cannot modify the elements of a tuple):

```
alpha = ('a','b','c')

# generating values/numbers at same time
for i,a in enumerate(alpha):
  print((i,a))
```

```
(0, 'a')
(1, 'b')
(2, 'c')
```

When one needs to loop over a dictionary, one may access the individual `keys` or `values` and loop over them using special methods for dictionaries. Or one can loop over them at the same time:

```
di = {'a': 0, 'b': 5, 'c': 10}

# get the keys
for k in di.keys():
  print(k)

# get the values
for v in di.values():
  print(v)

# can do both at the same time using items
for k,v in di.items():
  print((k,v))
```

```
a
b
c
0
5
10
('a', 0)
('b', 5)
('c', 10)
```

4.2 List Comprehension

Say you need to double the values in a list, this code will return a second list of the values doubled:

```
lv = [1,2,3]
ld = [] # make an empty list to start

for l in lv:
  ld.append(l*2) # append the doubled value to the ld list

print(ld)
```

```
[2, 4, 6]
```

An easier way to do this though is to use *list* comprehension. It is like a for loop, but on a single line:

```
lv = [1,2,3]
ld = [l*2 for l in lv]
print(ld)
```

```
[2, 4, 6]
```

A common pattern for list comprehensions is to filter a list:

```
lv = [1,2,3]
ld = [l*2 for l in lv if l < 3]
print(ld)
```

```
[2, 4]
```

You could of course do this example using the more typical `for` and `if` approach:

```
lv = [1,2,3]
ld = []

for l in lv:
  if l < 3:
    ld.append(l*2)

print(ld)
```

```
[2, 4]
```

But you can see how this saves 4 lines of code vs 1 in the list comprehension scenario.

In list comprehensions you can also do else statements, but they need to be *before* the for statement:

```
lv = [1,2,3]
ld = [l*2 if l < 3 else -1 for l in lv]
print(ld)
```

```
[2, 4, -1]
```

Even though these are called list comprehensions, you can do them for tuples or dictionaries as well.

```
lv = [1,2,3]

ld = {l:l**2 for l in lv}
print(ld) # dictionary
```

```
{1: 1, 2: 4, 3: 9}
```

Sometimes this is handy to swap a dictionaries keys and values. But remember, keys need to be unique and immutable.

```
ld = {'a': 1, 'b': 1, 'c': 2}

swap_dict = {v:k for k,v in ld.items()}

# note that there is no 1: 'a'!
# it was replaced with  1: 'b'
print(swap_dict)
```

```
{1: 'b', 2: 'c'}
```

In this example, since there are two values of 1 in the original `ld` dictionary, there can only be a single `1` key in `swap_dict`. Here the last value of `1: 'b'` wins, and replaces `1 : 'a'`.

Note if you have two lists you want to turn into a dictionary though, you don't need to loop at all. You can zip the two lists together, and convert that zipped object into a dictionary.

```
lv = [1,2,3]
keys = ['a','b','c']

ld = dict(zip(keys,lv))
print(ld)
```

```
{'a': 1, 'b': 2, 'c': 3}
```

4.3 Permutations, Combinations and Sets

So far we have been working with objects in which we define all the elements up front. Another common scenario is we need to define our own list or dictionary, built up from other objects. The itertools library in the python standard library has many common functions to create combinations and permutations of items. Here is an example of defining all the two based permutations of items in a list:

```python
import itertools

l1 = ['a','b','c','d']

for pairs in itertools.permutations(l1, 2):
  print(pairs)
```

```
('a', 'b')
('a', 'c')
('a', 'd')
('b', 'a')
('b', 'c')
('b', 'd')
('c', 'a')
('c', 'b')
('c', 'd')
('d', 'a')
('d', 'b')
('d', 'c')
```

> Note
>
> Python has a standard set of libraries that are always available, and itertools is one of them. If you install python, itertools will *always* be available.

Note that this has both `('a','b')` and `('b','a')`. However, you may only want single *combinations* invariant to the order.

```python
import itertools

l1 = ['a','b','c','d']

for pairs in itertools.combinations(l1, 2):
  print(pairs)
```

```
('a', 'b')
('a', 'c')
('a', 'd')
('b', 'c')
('b', 'd')
('c', 'd')
```

Another common example is looking at the product of two lists:

```
import itertools

l1 = ['a','b','c']
l2 = ['Z','Y']

for pairs in itertools.product(l1, l2):
  print(pairs)
```

```
('a', 'Z')
('a', 'Y')
('b', 'Z')
('b', 'Y')
('c', 'Z')
('c', 'Y')
```

One last common object I have not introduced so far are *sets*. This is useful if you have a list with repeated items, and you only want the unique ones:

```
l1 = ['a','b','a','d']
unique_letters = set(l1)
print(unique_letters)
```

```
{'d', 'a', 'b'}
```

Note that when you do `set(list)`, you do not necessarily get the items back in any particular order. So if the subsequent order of the items is important, you need to convert them to some other object:

```
l1 = ['a','b','a','d']
unique_letters = list(set(l1))
unique_letters.sort()
print(unique_letters)
```

```
['a', 'b', 'd']
```

You can do set operations, like union and difference. And you can see I define sets here in two different ways, one via coercing a list to a set, and the other using curly brackets but without colons like in a dictionary.

```
s1 = {'a','b','c'}
s2 = set(['a','d'])
```

```
su = s1.union(s2)
print(su) # union of sets

si = s1.intersection(s2)
print(si) # intersection of sets

sd = s1.difference(s2)
print(sd) # difference of sets

# note that difference is not symmetric
sd2 = s2.difference(s1)
print(sd2) # difference of second set from s1
```

```
{'c', 'd', 'a', 'b'}
{'a'}
{'c', 'b'}
{'d'}
```

Again because sets do not have an implicit order to them, if you run this code, the order of your output items may be different than above, but the items in the different set operations should be the same.

Note that sets can have variable object types within them. This example does not implicitly coerce the sub list to its individual items:

```
s = set(['a','b',('a','b')])
print(s)
```

```
{('a', 'b'), 'a', 'b'}
```

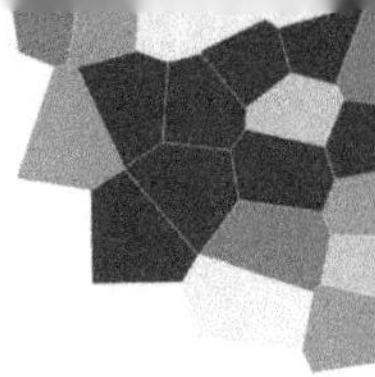

5 Functions and Libraries

These last 3 chapters have covered the basic objects and operations in the standard python library. Understanding these fundamentals are key to being a competent python developer. Real world projects will involve chaining together those prior operations I showed to conduct more complicated workflows.

This chapter will introduce you to functions and their utility. Lets continue with our example we used in the prior chapter on looping – you are a teacher and want to email your students a message. You could write code that contains the names of students, loop over the students, and email them one by one a pre-specified tailored message. Something like below for example:

```
student_emails = {'Joe Q': 'Person1@school.edu',
                  'Amelia Bedelia': 'Person2@school.edu',
                   ....,
                  'Billy Madison': 'Person20@school.edu'}

for name, email in student_emails.items():
  # custom message for each student
  message = f'''Hi {name}!, please note that homework
  assignment 3 is due tomorrow!
  cheers, your super rad teacher'''
  # logic to send email here, pretending this function
  # is defined already
  send_email(message,email)
```

This is ok to do a single time, but given the nature of the task, it is likely the teacher will want to do this multiple times. You could edit the script every time you wanted to send out a new email, but the easier solution is to *write a function*, that takes in inputs – like your dictionary of students and a custom message template. So then to update will be simpler:

```
def email_reminder(student_emails,custom_message):
  for name, email in student_emails:
    message = f"{custom_message}".format(name=name)
    send_email(message,email)
```

And then when you want to send a new email, your script to update is much simpler, such as:

```
# new custom email message
custom_message = """Hey {name}, the final is at 8 am.
Be there or be square. Andy Wheeler"""

# get student emails from external file (pandas function)
student_emails = read_csv('PythonClass_EmailList.csv')

# turn into dictionary like function expects
student_emails = student_emails.to_dict(orient='records')

# send email reminder
email_reminder(student_emails,custom_message)
```

I have done a few shortcuts in these examples to make them simpler. I assumed functions `send_email` and `read_csv` are available. The `read_csv` function is available in the `pandas` library. I will show later in this chapter how to install and import different libraries into your python script. The `send_email` function is not a generally available function, but you can write it yourself if you want, see https://andrewpwheeler.com/2015/03/02/emailing-with-python-and-spss/ for an example! Generally more complicated scripts will take writing several functions to make the final script simple and easy to update over time.

This chapter has four sections. The first section shows how to define *your own* functions and pass in arguments. The second section shows how to save your own functions in a separate scripts file, and then how to *import* those functions into your current session. The third section shows how to install different libraries into your python environment, and then load them into your current script. It also gives a brief breakdown of understanding versions of libraries, and creating python environments. The fourth section gives a very brief introduction to the distinction between class methods and functions in python.

I mention "brief" for several of these sections (python environments and introducing classes) as they could be greatly expanded. I cover what I believe is necessary though for introductory material (for more advanced data and project management, I have plans to develop an additional book).

5.1 Defining your own functions

To define your own function in python, you use the format `def function_name(arguments):`. Similar to if statements or for loops, you then use whitespace to determine the operations that happen *inside* your function. At the end of your function, you probably want to have it `return` some specific result. It is easier to show than to say, this simple example takes an input, `a`, divides `a` by 2, and returns that result:

```
# example of defining your own function
def my_func(a):
  res = a/2
  return res
```

```
# now you can call that function you defined
mv = my_func(4)
print(mv)
```

```
2.0
```

In this example, `a` is an *argument* to the function. Functions can have multiple arguments, this example has two:

```
# example with two inputs
def my_func(a,b):
  res = (a + 1)/(b + 1)
  return res

mv = my_func(1,2)
print(mv)
```

```
0.6666666666666666
```

In this example, python knows that the first argument to `my_func(1,2)` is meant to be `a`, and assigns the value of 1 to `a` *inside* that function. It then knows the second argument is `b`, and assigns it the value of 2. You can however explicitly name the arguments, what are called keyword arguments in python. When you use keyword arguments, the order does not matter.

```
# example of defining your own function
def my_func(a,b):
  res = (a + 1)/(b + 1)
  return res

# used named arguments, not implicit
mv = my_func(b=1,a=2)
print(mv)
```

```
1.5
```

You can assign arguments *default* values. Here, if I do not pass in a `b` argument, it defaults to `b` having a value of 3.

```
# can give default arguments
def my_func(a,b=3):
```

```
    res = (a + 1)/(b + 1)
    return res

# With default argument for b
mv = my_func(a=1)
print(mv)
```

```
0.5
```

Note however that default values always need to be defined on the `def` line after non-default arguments, so `def my_func(a=1,b):` would return an error. If you give all values defaults, you can then call the function empty without passing in any arguments, and get a default returned value:

```
# can give default arguments
def my_func(a=0,b=3):
    res = (a + 1)/(b + 1)
    return res

# With default argument for both
# just empty call
mv = my_func()
print(mv)
```

```
0.25
```

These examples I have passed in arguments or keyword arguments (sometimes referred to in python slang as args or kwargs) one by one. If you have an object that has the elements you want to pass to the function though, such as a dictionary (for kwargs) or a list (for args), you can pass those to the function all at once. This is a special python operator, using `**dict` to unpack kwargs, and `*list` to unpack args.

```
# example of defining your own function
def my_func(a,b):
    res = (a + 1)/(b + 1)
    return res

# can pass in dictionary mapping
# for keyword args
arg_di = {'a': 1, 'b': 3}
mv = my_func(**arg_di)
print(mv)
```

```
# or if just in order
# can unpack list/tuple
arg_li = [2,4]
mv2 = my_func(*arg_li)
print(mv2)
```

```
0.5
0.6
```

It may be the case though that for your function, it is easier to have a single argument itself be some type of python container, such as a list. You can access elements of that argument same as I showed in Chapter 2:

```
# can pass in different objects
def my_func(val):
  res = (val[0] + 1)/(val[1] + 1)
  return res

# now expects a list or tuple
mv = my_func([1,2])
print(mv)
```

```
0.6666666666666666
```

One convenient trick with python functions, you can return *multiple* values. Here I show an example returning two values using tuple unpacking.

```
# can return multiple objects
def my_func(a,b):
  r1 = (a + 1)/(b + 1)
  r2 = min(a,b)
  return r1, r2

# returns two results
m1, m2 = my_func(1,2)
print(m1, m2)
```

```
0.6666666666666666 1
```

When python evaluates a function, it stops whenever the first return occurs. Here I write the function to return `None` to prevent dividing by zero:

```
# the first return stops the function
def my_func(a,b):
  if (b+1) == 0:
    print('r1 is undefined, divided by zero')
    return None
  # Note this is outside of if
  r1 = (a + 1)/(b + 1)
  return r1

# should return None
m1 = my_func(-2,-1)
print(m1)
```

```
r1 is undefined, divided by zero
None
```

One aspect I have not discussed so far is local scope vs global scope. Inside of a function, you can access objects outside of the function. But certain actions are only done locally inside of a function. Here I define `a` outside of the function, and use that `a` object inside the function:

```
# this is global, but is accessible
# inside the function
a = 6

# note I don't pass a anymore as
# an argument
def my_func(b):
  r1 = (a + 1)/(b + 1)
  return r1

# this should return 1
m1 = my_func(6)
print(a,m1)

# but can change a in global
# and now will return 2.5
a = 4
m2 = my_func(1)
print(a,m2)
```

```
6 1.0
4 2.5
```

If you had your original function with an a argument though, that a object has only a local scope inside of the function. It effectively ignores the a object in the global scope.

```
a = 6 # this is global

def my_func(a,b):
  # but a inside here a is local
  r1 = (a + 1)/(b + 1)
  return r1

# this should return 0.5, not 1
m1 = my_func(2,5)

# note however a in global is still
# the same value, it has not changed
print(a,m1)
```

```
6 0.5
```

Although it is common to use global objects inside of functions, you often do not want to modify objects inside of a function. Repeated calls to the function can have unintended consequences. Consider this example, that appends a value to a list. It is probably not the intended behavior (there is no point in `return` if the intended behavior is to just append values to a single list):

```
x = [] # x is global outside

def xa(a):
  x.append(a/2)
  return x

# This modifies the global list
x1 = xa(1)
print(x)
print(x1)

# This even modifies x1
x2 = xa(2)
print(x)
print(x1)
print(x2)
```

```
[0.5]
[0.5]
[0.5, 1.0]
[0.5, 1.0]
[0.5, 1.0]
```

On occasion however, you do want to modify some global object. This is an example function that does not return any value, but is intended to modify a global dictionary that has some min/max statistics:

```
# default values for dictionary
di = {'min': 0, 'max': 10}

# function that updates min/max
# given an argument
def update_di(x):
  if x < di['min']:
    di['min'] = x
  elif x > di['max']:
    di['max'] = x
  # no need for return

update_di(-1)
print(di) # min is updated

update_di(11)
print(di) # max is updated
```

```
{'min': -1, 'max': 10}
{'min': -1, 'max': 11}
```

If you have a need for something like this, but *do not* want to modify the initial object, you can use the `copy` method to return a copy of the original object, but it is not linked.

```
# default values for dictionary
di = {'min': 0, 'max': 10}

# function that updates min/max
# given an argument
def update_di(x):
    di2 = di.copy() #copied
    if x < di['min']:
      di2['min'] = x
    elif x > di['max']:
      di2['max'] = x
```

```
    return di2

dr = update_di(-1)
print(di) # this is the same
print(dr) # this is updated
```

```
{'min': 0, 'max': 10}
{'min': -1, 'max': 10}
```

This example shows how to copy a dictionary, but you can use copy on lists (or many other python objects) as well.

One final example I want to show about functions, when you write functions, it is good to have documentation (or "docstrings" as they are often referred to in python). To check out the documentation for a function, you can type `help(function)`, and it prints the documentation to the terminal. Here for example is the documentation for the `sum` function:

```
# printing the docs for
# the sum func
help(sum)
```

```
Help on function sum in module __main__:

sum(x)
    Return the sum of a 'start' value (default: 0)
    plus an iterable of numbers

    When the iterable is empty, return the start value.
    This function is intended specifically for use with
    numeric values and may reject non-numeric types.
```

If you place text in triple quotes below a function, it will print that out when running `help(my_function)`. There are different ways to format docstrings so they can be auto compiled into nice documentation, but just having a few notes is often sufficient to help future readers of your code (including yourself). Often the notes include a description of what the function does, what the arguments refer to, and what the function returns.

```
# example with two inputs
def my_func(a,b):
    """
    This function returns
```

```
    (a + 1)/(b + 1)

    a and b should be numeric values
    b should not be -1!
    """
    res = (a + 1)/(b + 1)
    return res

# see the print out
help(my_func)
```

```
Help on function my_func in module __main__:

my_func(a, b)
    This function returns

    (a + 1)/(b + 1)

    a and b should be numeric values
    b should not be -1!
```

5.2 Creating your own functions in a seperate file

Similar to Chapter 1, now I am going to have you make a seperate set of `.py` files that contain functions. The reason for this is that imagine you have a project that has 20 functions, you do not want to keep them all at the beginning of your script. It clutters what you want to do and makes it more difficult for others to understand what the script is meant to accomplish.

Also imagine a project in which you want to re-use functions. Say you have a consistent set of functions to query your local database, and generate crime statistics given an input date range and crime types. In that scenario, you should create a location that has your functions you want to use multiple times, and *import* them into your script that can change over time. This is because you will ultimately edit that function over time, and so having it in a single place allows you to modify it and it will update for all of your scripts.

So here I am going to show how to import different files. These examples are intentionally very simple sets of functions – the point is to understand how to store functions in consistent locations so you can use them across multiple scripts.

First, create a new folder, here I created a folder `B:\func_example`. Then within that folder create a new folder named `src` (short for "source"). Now within the `src` folder, create two files. `func1.py` and `func2.py`. The contents of `func1.py` should be:

```
# This is the function in the file
# func1.py

def my_function(x):
  print('This is function 1')
  return x + 1
```

And the contents of `func2.py` should be:

```
# This is the file func2.py

from . import func1

def my_function2(x):
  print('This is function 2')
  return x + 2
```

Once finished, your `src` folder should look like this, with two python files:

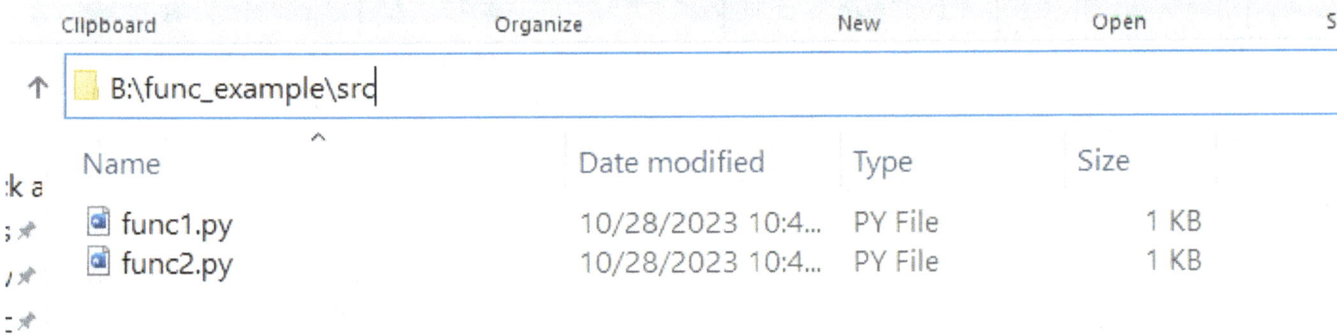

Now in your terminal, I am going to show how to import these functions. But first, you need to navigate the terminal so you are at the *root* of the project. In this screenshot, this is at `B:\func_example` – but on your machine this location may be different! You should be one folder up from the `src` folder where you placed your functions. Then you can enter into the REPL by typing the command `python`. Once you are at that stage, type the command `from src import func1`. Now you have your function available, and can run `func1.my_function(0)` and see the result:

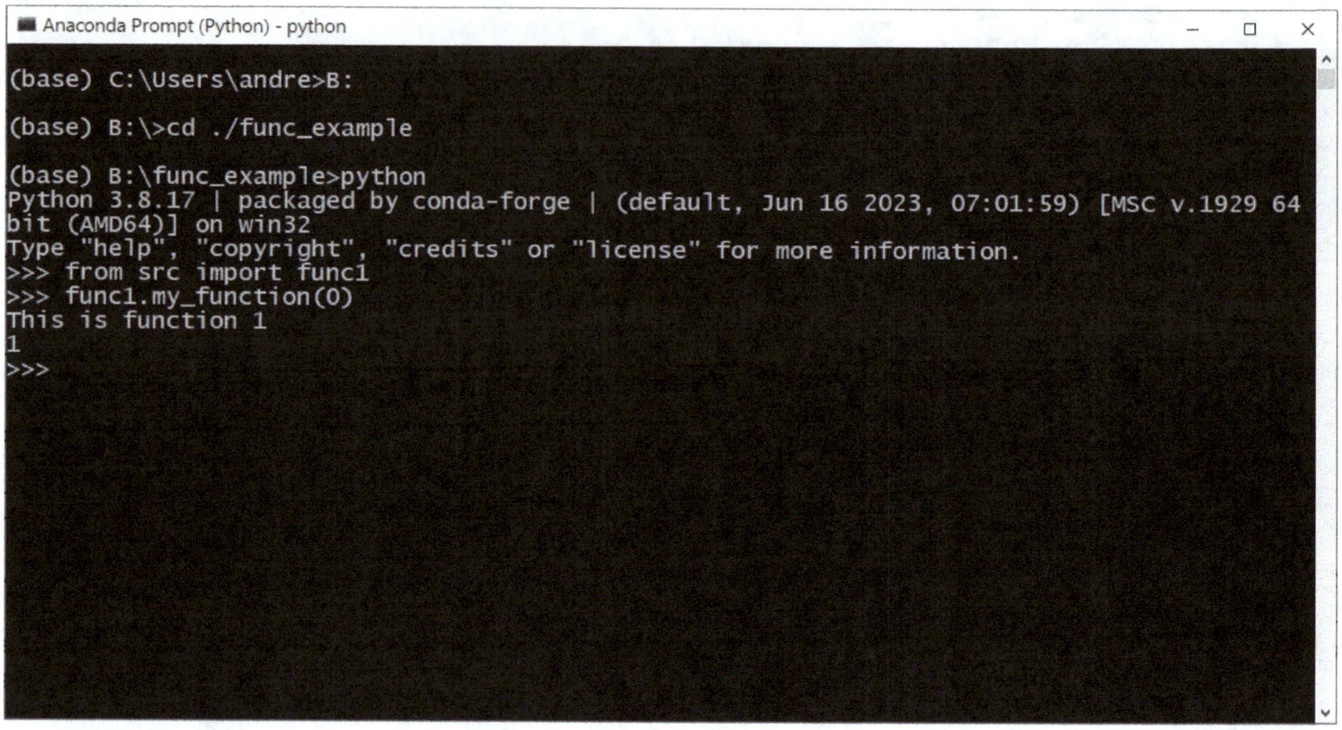

```
Anaconda Prompt (Python) - python

(base) C:\Users\andre>B:

(base) B:\>cd ./func_example

(base) B:\func_example>python
Python 3.8.17 | packaged by conda-forge | (default, Jun 16 2023, 07:01:59) [MSC v.1929 64
bit (AMD64)] on win32
Type "help", "copyright", "credits" or "license" for more information.
>>> from src import func1
>>> func1.my_function(0)
This is function 1
1
>>>
```

The reason this works is that python looks for files along its path. By default, python adds where it is being run to the path – the root folder we are currently in. You can do different things to either modify where python is being run, e.g. by changing the current directory (e.g. `import os` and then `os.chdir(r'B:\func_example')`), or by adding where the files are to the path (e.g. `import sys` and then `sys.path.append(r'B:\func_example')`).

But more professional software engineering typically fixes the location scripts are run, and does not modify those external elements. (You may append *the system* path in Windows if you have scripts you use across many projects, so you do not always need to type `sys.path.append` at the start of every python script.)

We can also import *specific* functions. So for example, we can do

```
from src.func2 import my_function2
```

```
Anaconda Prompt (Python) - python

(base) B:\func_example>python
Python 3.8.17 | packaged by conda-forge | (default, Jun 16 2023, 07:01:59) [MSC v.1929 64 bit (AMD64)] on win32
Type "help", "copyright", "credits" or "license" for more information.
>>> from src import func1
>>> func1.my_function(0)
This is function 1
1
>>> from src.func2 import my_function2
>>> my_function2(3)
This is function 2
5
>>>
```

So if you only need one specific function, you may just import what you want/need in your script to make it simpler. Note that you can *rename* functions on import, here I will change the name of the custom function to something much shorter, `from src.func2 import my_function2 as f2`

```
Anaconda Prompt (Python) - python

(base) B:\func_example>python
Python 3.8.17 | packaged by conda-forge | (default, Jun 16 2023, 07:01:59) [MSC v.1929 64 bit (AMD64)] on win32
Type "help", "copyright", "credits" or "license" for more information.
>>> from src import func1
>>> func1.my_function(0)
This is function 1
1
>>> from src.func2 import my_function2
>>> my_function2(3)
This is function 2
5
>>> from src.func2 import my_function2 as f2
>>> f2(3)
This is function 2
5
>>>
```

This can make long function names easier to write – a very nice feature if the function names are very long (and you need to call them multiple times in a script).

> Note
>
> You can import *all* functions from a particular library via `from library import *`. This is not typically recommended though, as it makes it more difficult to know the provenance of any particular function. For example, the library could have its own internal `sum` function, and it would *overwrite* the standard python libraries sum function. This would make it very difficult to debug the code.

Note in our original `func2.py` python script that contained our function, we had the line

```
from . import func1
```

This is a *relative* import, so you can have functions in the same folder import functions defined in other scripts. To make it clear the implications of this, go ahead and `exit()` the current REPL session. Then start a new session (by typing `python` at the terminal), and then once in a new REPL session type `from src import func2 as f2`.

```
Anaconda Prompt (Python) - python
5
>>> from src.func2 import my_function2 as f2
>>> f2(3)
This is function 2
5
>>> exit()

(base) B:\func_example>python
Python 3.8.17 | packaged by conda-forge | (default, Jun 16 2023, 07:01:59) [MSC v.1929 64
bit (AMD64)] on win32
Type "help", "copyright", "credits" or "license" for more information.
>>> from src import func2 as f2
>>>
```

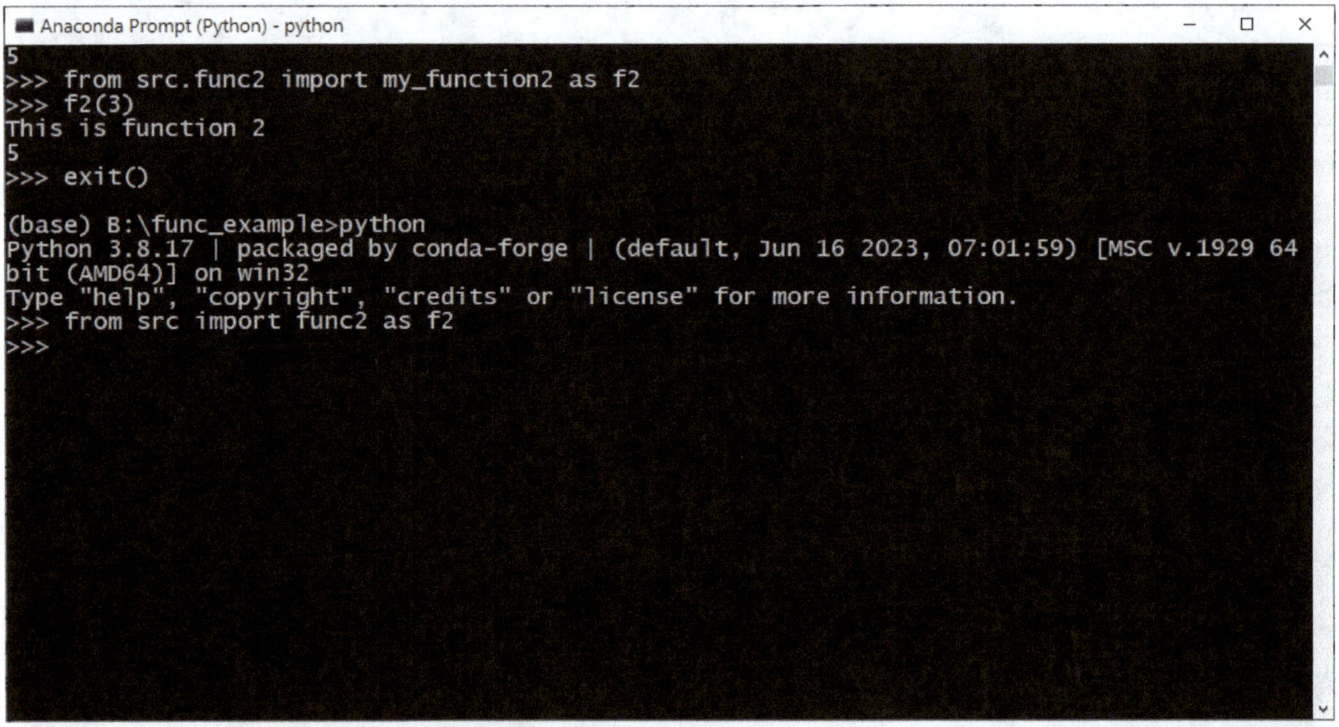

Now we can call `f2.func1.my_function`! Even though we did not import `func1` directly into the REPL, it was imported via `func2`:

```
Anaconda Prompt (Python) - python
(base) B:\func_example>python
Python 3.8.17 | packaged by conda-forge | (default, Jun 16 2023, 07:01:59) [MSC v.1929 64
bit (AMD64)] on win32
Type "help", "copyright", "credits" or "license" for more information.
>>> from src import func2 as f2
>>> f2.func1.my_function(4)
This is function 1
5
>>>
```

When importing libraries there are two more common tricks I want to show. First, if you are importing multiple functions from a library, you can import them on the same line. So something like `from library import somefunc1, somefunc2`. So with these two test functions, an example could be: `from src.func2 import func1, my_function2`.

```
Anaconda Prompt (Python) - python
(base) B:\func_example>python
Python 3.8.17 | packaged by conda-forge | (default, Jun 16 2023, 07:01:59) [MSC v.1929 64
bit (AMD64)] on win32
Type "help", "copyright", "credits" or "license" for more information.
>>> from src.func2 import func1, my_function2
>>> func1.my_function(1)
This is function 1
2
>>> my_function2(2)
This is function 2
4
>>>
```

Second, when you are interactively rewriting functions, when importing from a different location, the functions are by default not edited. So first, exit and restart a new REPL session, and then do `from src import func2 as f2`. Then run `f2.my_function2(3)`.

But say we were in the process of editing our function, and we wanted to change it some. You could exit and then restart a REPL session, but an easier trick is to use the `importlib` library and its `reload` function.

So say we edit `func2.py` to now be:

```
# This is the file func2.py

from . import func1 as f1

def my_function2(x):
  print('Yo this is function 2 updated')
  return x + 3
```

And now back in the REPL, type in `reload(f2)` (we are reloading the *name* we used for the module). Now you can re-run `f2.my_function2(3)` and it will show the updated output. Here is a screenshot of those steps to follow along:

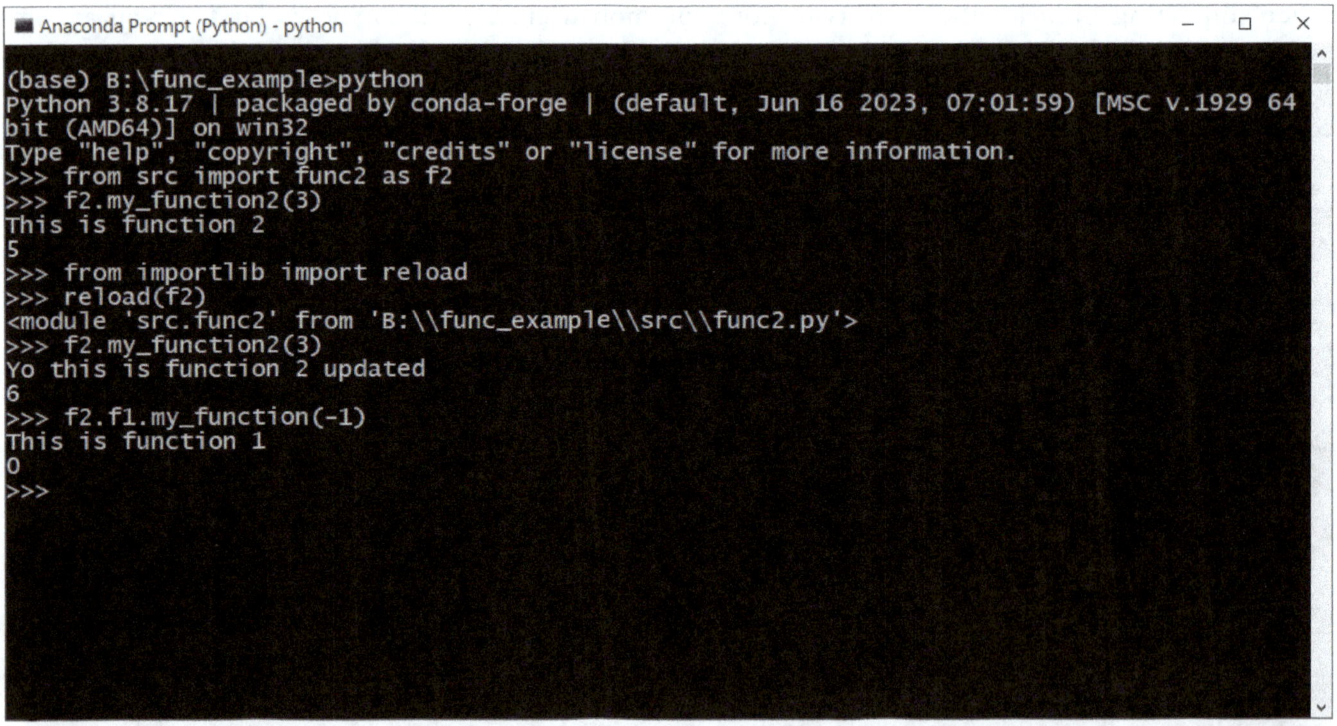

```
Anaconda Prompt (Python) - python

(base) B:\func_example>python
Python 3.8.17 | packaged by conda-forge | (default, Jun 16 2023, 07:01:59) [MSC v.1929 64
bit (AMD64)] on win32
Type "help", "copyright", "credits" or "license" for more information.
>>> from src import func2 as f2
>>> f2.my_function2(3)
This is function 2
5
>>> from importlib import reload
>>> reload(f2)
<module 'src.func2' from 'B:\\func_example\\src\\func2.py'>
>>> f2.my_function2(3)
Yo this is function 2 updated
6
>>> f2.f1.my_function(-1)
This is function 1
0
>>>
```

5.3 Installing other libraries

This chapter so far has shown examples of writing your own functions. In prior chapters I have also shown libraries that are available by default in all of python, such as `itertools` and `re`. Both of those packages are in the standard python library, so you will never need to install those libraries.

One of the benefits of open source software is that many individuals contribute open source packages. If someone else has already written the function, you do not need to write it yourself, you can install their package and use those functions in your code. Installing other packages also has the benefit that the authors have often done vetting of their own – such as writing unit tests to make sure the functions behave as they should.

> Note
>
> One does need to be wary though of installing public packages. For example, a common type of fraud with open source packages is *typo-squatting*. Say someone wants to install the `pandas` library, but instead types `pip install panda` (note the lack of an `s` at the end). Someone may upload a package named `panda`, that has all the same functions as the original pandas library (so the user does not even realize there is a problem), but also has functions that send potentially sensitive user data (such as passwords saved on your local system) to the hacker. Or runs some other type of malicious program when first installing the package.

One thing to note though when installing new packages, you often want to make a *new* python environment. This is because when installing packages, the local system needs to deal with conflicts. Say you want to install `packageA`, and `packageA` depends on `packageZ` being greater than version 3. But you have another `packageB`, and `packageB` needs `packageZ` to be less than version 3. In this scenario, you cannot have both `packageA` and `packageB` installed in the same environment.

Pure environment conflicts do not happen often, but dealing with varying package inter-dependencies does occur frequently, and installing a new package can commonly have effects on other packages that are difficult to foresee. So it is a good practice to create a new environment for each python project that you need to install new packages.

Here I am going to show how to handle environments in the `conda` package manager. The full anaconda distribution of python is intended specifically for people who do data analysis, and comes installed with many of the packages (such as `pandas` for working with tabular datasets). I will show, through screenshots of the terminal, how to create a new environment and install packages.

First, from the conda terminal, type

```
conda create --name panda_env python=3.10 pip pandas
```

This will create a new environment, named `panda_env`. This installs the 3.10 version of python, and then adds in the `pip` package manager, as well as the `pandas` library.

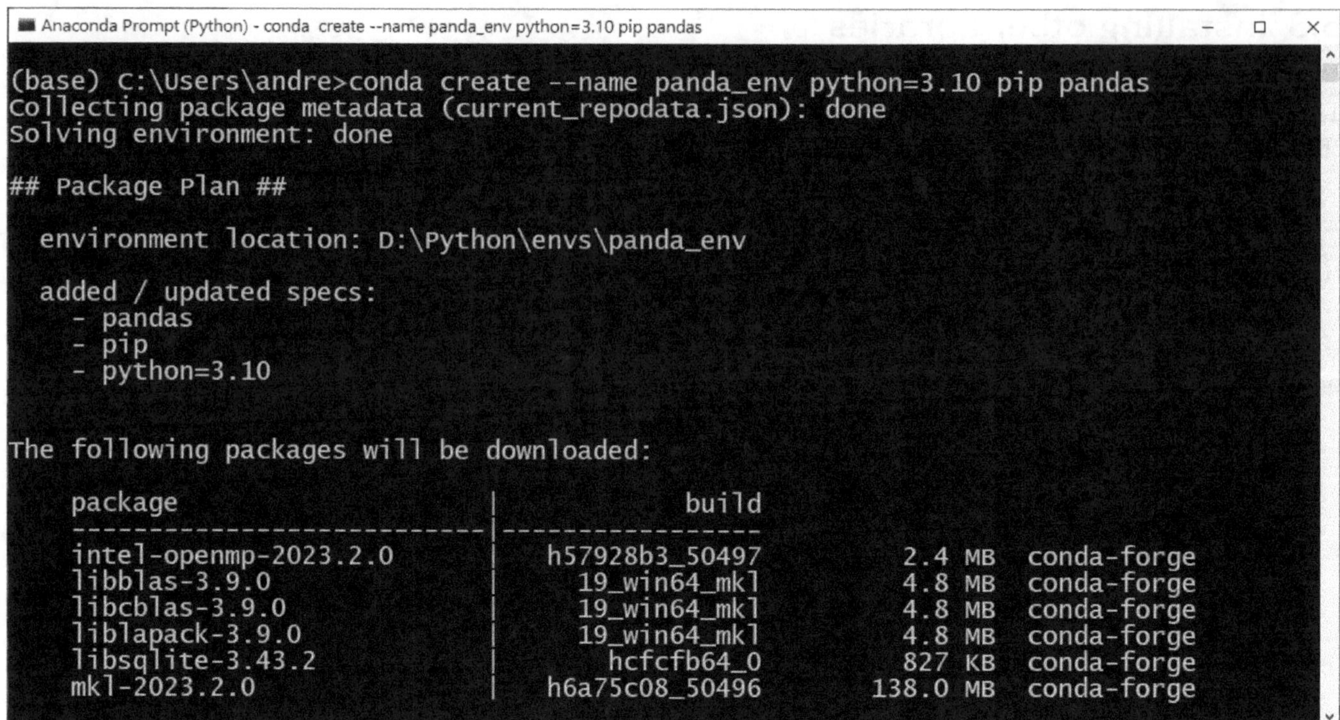

You will see that there are many packages that also need to be installed, not just pandas and pip! (There are even more than can be shown in this screenshot.) This is common with more advanced libraries, they have many inter-dependencies. But conda figures that all out for us.

Now it will ask `Proceed ([y]/n)?`, and if you type in `y` into the terminal it will install that new environment. Then to *activate* that new environment, at your terminal type `conda activate panda_env`.

```
Anaconda Prompt (Python)

Proceed ([y]/n)? y

Downloading and Extracting Packages:

Preparing transaction: done
Verifying transaction: done
Executing transaction: done
#
# To activate this environment, use
#
#     $ conda activate panda_env
#
# To deactivate an active environment, use
#
#     $ conda deactivate

(base) C:\Users\andre>conda activate panda_env

(panda_env) C:\Users\andre>
```

Now lets say we want to also include the newest version of the `scipy` library, version `1.11.3` at the time of this writing. We can install this library either via `conda install scipy==1.11.3` or `pip install scipy==1.11.3`. I prefer the later (so use conda to create new python environments, but then have pip deal with the dependencies), but there are some situations where it is easier to have conda deal with all of the installs. (Sometimes when installing complicated dependencies on Windows, it is easier to install the tougher packages using conda first. This is how I typically handle packages such as `pytorch` (for deep learning) or `geopandas` for geographic data manipulation.)

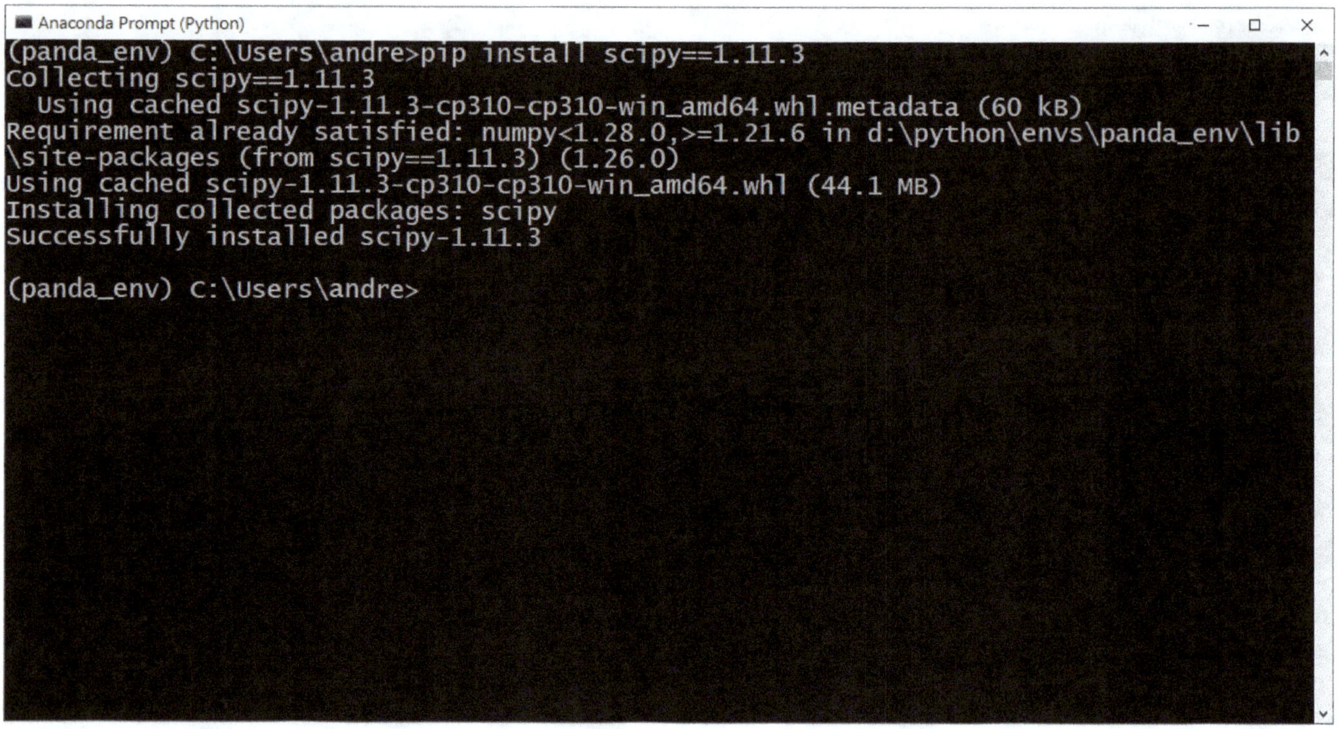

```
Anaconda Prompt (Python)
(panda_env) C:\Users\andre>pip install scipy==1.11.3
Collecting scipy==1.11.3
  Using cached scipy-1.11.3-cp310-cp310-win_amd64.whl.metadata (60 kB)
Requirement already satisfied: numpy<1.28.0,>=1.21.6 in d:\python\envs\panda_env\lib
\site-packages (from scipy==1.11.3) (1.26.0)
Using cached scipy-1.11.3-cp310-cp310-win_amd64.whl (44.1 MB)
Installing collected packages: scipy
Successfully installed scipy-1.11.3

(panda_env) C:\Users\andre>
```

Here we use `==` in the install to force installing a certain version. You could also specify greater than a specific version as well, e.g. `pip install scipy>=1.11`.

> Note
>
> The way typical packaging versioning works, it can be read as `major.minor.fix`, so scipy 1.11.3 can be read as "major version 1, minor version 11, update 3". There are no rules (package authors can use whatever versioning scheme they want), but in general updates are for bug fixes, minor versions add in new functions, and major version upgrades are typically reserved for potentially code breaking changes. So updating from `1.10` to `1.11` probably will not break your code (with version 1.11 being newer than version 1.10).

If you have many packages you want to install in a particular environment, a good practice is to have a file, `requirements.txt`, in the root of your project that lists all of the different packages. Here for example I include `networkx` (a python library for network analysis) and `matplotlib` (a library to make graphs), along with scipy, in the requirements.txt file:

```
# This is a requirements.txt file
# I first created the environment using
# conda create --name panda_env python=3.10 pip pandas
scipy==1.11.3
networkx
matplotlib
```

Here you can see I even include comments in the requirements.txt for creating the initial conda environment. This allows others to easily replicate the steps you used to create the environment. Now if you run `pip install -r requirements.txt` it will install these new libraries (and skip over scipy, since that is already installed):

```
Anaconda Prompt (Python)

(panda_env) B:\func_example>pip install -r requirements.txt
Requirement already satisfied: scipy==1.11.3 in d:\python\envs\panda_env\lib\site-pa
ckages (from -r requirements.txt (line 4)) (1.11.3)
Collecting networkx (from -r requirements.txt (line 5))
  Downloading networkx-3.2.1-py3-none-any.whl.metadata (5.2 kB)
Collecting matplotlib (from -r requirements.txt (line 6))
  Downloading matplotlib-3.8.0-cp310-cp310-win_amd64.whl.metadata (5.9 kB)
Requirement already satisfied: numpy<1.28.0,>=1.21.6 in d:\python\envs\panda_env\lib
\site-packages (from scipy==1.11.3->-r requirements.txt (line 4)) (1.26.0)
Collecting contourpy>=1.0.1 (from matplotlib->-r requirements.txt (line 6))
  Downloading contourpy-1.1.1-cp310-cp310-win_amd64.whl.metadata (5.9 kB)
Collecting cycler>=0.10 (from matplotlib->-r requirements.txt (line 6))
  Downloading cycler-0.12.1-py3-none-any.whl.metadata (3.8 kB)
Collecting fonttools>=4.22.0 (from matplotlib->-r requirements.txt (line 6))
  Downloading fonttools-4.43.1-cp310-cp310-win_amd64.whl.metadata (155 kB)
     ---------------------------------------- 155.5/155.5 kB 849.7 kB/s eta 0:00:00
Collecting kiwisolver>=1.0.1 (from matplotlib->-r requirements.txt (line 6))
  Downloading kiwisolver-1.4.5-cp310-cp310-win_amd64.whl.metadata (6.5 kB)
Collecting packaging>=20.0 (from matplotlib->-r requirements.txt (line 6))
  Downloading packaging-23.2-py3-none-any.whl.metadata (3.2 kB)
Collecting pillow>=6.2.0 (from matplotlib->-r requirements.txt (line 6))
  Downloading Pillow-10.1.0-cp310-cp310-win_amd64.whl.metadata (9.6 kB)
Collecting pyparsing>=2.3.1 (from matplotlib->-r requirements.txt (line 6))
  Downloading pyparsing-3.1.1-py3-none-any.whl.metadata (5.1 kB)
```

When first creating an environment, it is easier to only specify the versions for particular important packages, and then let the package manager figure out all of the interdependencies for the packages. Once your environment is set up though, you can use `pip list` to show the packages and their explicit versions:

```
(panda_env) B:\func_example>pip list
Package         Version
--------------- -----------
contourpy       1.1.1
cycler          0.12.1
fonttools       4.43.1
kiwisolver      1.4.5
matplotlib      3.8.0
networkx        3.2.1
numpy           1.26.0
packaging       23.2
pandas          2.1.2
Pillow          10.1.0
pip             23.3.1
pyparsing       3.1.1
python-dateutil 2.8.2
pytz            2023.3.post1
scipy           1.11.3
setuptools      68.2.2
six             1.16.0
tzdata          2023.3
wheel           0.41.2

(panda_env) B:\func_example>
```

Or an easier way to save a nicely formatted file is to use `pip list --format freeze`. You can also *pipe* commands to an output file, e.g.

```
pip list --format freeze > req_freeze.txt
```

This gives you *exact* versions in a way that `pip install -r req_freeze.txt` in the future will install the exact same library versions as you currently have in your environment.

5.4 Object Methods vs Functions

This chapter I have shown you how to write your own functions, and in Chapter Two I showed working with various python objects. One thing I have avoided so far though is the distinction between functions and objects. Python is what is called an *object-oriented* programming language. In short, *everything* in python is an object (including functions).

Some beginner texts have more details on the object oriented nature of python – I don't think for those getting started in data analysis though it matters all that much. So I only note it briefly here as something "good to know" – applications you want to use it for will probably only come up when you grow from a beginner to a more advanced python programmer.

Here I am going to create a simple new type of object in python, a `Point` object. The reason I am showing this is so you can have a bit more understanding on the difference between functions and methods, as those first exposed to python can sometimes be confused about `function(args)` vs `object.method(args)`. I hope by giving an example of creating your own object, it becomes a bit clearer (although again most beginner applications will not have a need to write their own methods).

The Point object I create here has two *attributes* – an x coordinate and a y coordinate. I also give that object a few methods; a distance method (to calculate the Euclidean distance from a new set of x/y coordinates), and a move method (that modifies the Point objects internal X/Y coordinates).

The way I think about it is that objects have data associated with them, as well as special functions (called methods) that operate on that data internal to the object.

```
# need to import some math funcs
from math import radians, sin, cos

# Making my own Point object
class Point:
  def __init__(self,x,y):
    self.x = x
    self.y = y
  def distance(self,x2,y2):
    """
    Distance between Point
    and new x/y coords
    """
    xd = (self.x - x2)**2
    yd = (self.y - y2)**2
    euc_dist = (xd + yd)**(1/2)
    return euc_dist
  def move(self,dist,orient):
    """
    Updates the Point object
    x/y to new locations
    given distance and orientation
    in degrees
    """
    rads = radians(orient)
    self.x = self.x + sin(rads)*dist
    self.y = self.y + cos(rads)*dist
```

You can see we have several new things going on compared to the prior just defining our own functions. First, I introduce a new `class Name:`. This is python's idiosyncratic way of defining a new *class* for an object. Second, underneath the class object we have several functions (started with a `def` statement). First, we have a special `def __init__` – this is how an object is *initialized*. For our Point object, we pass it in two coordinates, an x coordinate and a y coordinate. These are assigned to the object via the `self.x = x` (and ditto for the y coordinate). These again are called the *attributes* of the object. Classes you define can have as many attributes and as many methods as you want. Remember for indentation rules, that each method needs to be indented the same amount, and for code underneath each method it needs to follow further indentation rules.

> Note
>
> In python objects, the functions that start with *two* underscores are called "dunder" methods. Most objects have a consistent set of dunder methods, such as `__init__` and `__repr__` for example (the latter is what gets printed to the console when you print the object).

The function definition of `function(self,args)` is a special way to define methods for objects. Self here means the function can access the objects attributes in the objects methods. So we have a function to define the distance to a set of new x/y coordinates given the objects internally defined X/Y coordinates. It references the objects own x/y coordinates, via `self.x` and `self.y` respectively.

The second object method `move`, here *modifies* the internal Point objects x and y coordinates given distance and orientation. So the method does not return any values, but modifies the Point objects internal x/y attributes.

```
# Using the point object
# defined in prior code cell

# defining a new point
P1 = Point(1,1)

# distance to up one unit
print(P1.distance(1,2))

# can check out the attributes
print(P1.x,P1.y)

# can modify the attributes
P1.move(1,0)
print(P1.x,P1.y) # P1 is updated
```

```
1.0
1 1
1.0 2.0
```

Note that similar to defining docstrings with functions earlier, you can print out an objects methods via `dir(object)`. And you can treat `object.method` the same as a function, and print out `help(object.method)`.

```
# Can print methods/help
# same as functions previously

# defining a new point
P1 = Point(1,1)
```

```
# can print out the objects methods
print(dir(P1)[-4:])

# can print help for method
print(help(P1.distance))
```

```
['distance', 'move', 'x', 'y']
Help on method distance in module __main__:

distance(x2, y2) method of __main__.Point instance
    Distance between Point
    and new x/y coords

None
```

I present this example to illustrate how creating your own objects and methods work under the hood – methods are just special functions that leverage an objects attributes. They can return new values (distance here), or can modify the objects attributes (move here).

> Note
>
> I do not show them here, but a convention with python is that methods with a single underscore, e.g. `def _method(self,args):` is meant to signal that the method is a "private" method. I use square quotes with private here, as python doesn't really have private methods in the sense that users cannot call them (like in javascript). It is just a signal that people are likely not meant to call them directly.

The Point object I created here is very simple. It could be extended, but I use it merely as a way to illustrate the difference between functions and object methods. Although crime analysts use GIS analysis, one of the reasons I say at the beginning of this section that using object-oriented programming is more advanced material, most data applications will not be working with a single set of x,y coordinates, but a tabular dataset of many x,y coordinates. Say for example burglaries in a jurisdiction geocoded to the address, or a set of ATM locations as crime generators.

In those examples, it makes more sense to use objects and functions in other packages, in particular numpy and pandas, and do operations on vectors of data. That is what the following chapter goes into detail, getting started with tabular data analysis.

6 Working with Tabular Data

So far I have shown operations on python objects, such as lists and dictionaries. Individuals who conduct *data analysis* though tend to work with matrices of data, and want to conduct *mathematical* operations on those objects. For example, say you did a hotspots policing experiment, and had a list of recorded crimes in the treated and control areas each week – so something like `treated = [1,6,8,0,10]` and `control = [5,2,8,12,9]`. You may want to look at metrics at the difference in treated vs control (if the hotspots intervention is effective, treated area will have fewer crimes on average).

These mathematical operations are much easier to accomplish using two python libraries – numpy and pandas. The numpy library is for vectors and matrices of (mostly numeric) data. Pandas is more general (and is built on top of numpy) – pandas matrices can contain mixed data (data with different types, e.g. some numbers, some dates, some strings).

This chapter will provide many examples of doing data operations on these two types of objects. You will probably not memorize the entire set of functions for either numpy or pandas – I commonly google both when I am doing a project. The chapter does not illustrate all of the potential numpy and pandas functions/methods either.

But in general knowing that a certain operation is possible is the first step to learning – and this chapter provides a reference point for the types of common data manipulations data scientists encounter in their regular tasks.

> Note
>
> The code examples in this chapter require that the libraries `numpy` and `pandas` are installed. If you do not have them installed in your python environment, you can check out Chapter 5 to see how to install those libraries. If you are using the anaconda distribution of python, they are likely installed in the default distribution though.

6.1 numpy

First, I will show conducting mathematical operations using the *numpy* library. The main convention when writing numpy code when importing is to abbreviate the library to `np`. Here I create a *vector* of integers.

```
import numpy as np
```

```
x = np.array([1,5,2])
x
```

```
array([1, 5, 2])
```

This offhand is not much different than a list of integers, but numpy arrays are special, and we can conduct mathematical operations on the array in *vectorized* form. This example shows you can conduct math operations on the array and a *scaler* constant value.

```
# Will omit the import statement
# from here on
x = np.array([1,5,2])

print(x - 1)
print(x + 1)
print(x/2)
print(x*2)
print(x**2)
```

```
[0 4 1]
[2 6 3]
[0.5 2.5 1. ]
[ 2 10  4]
[ 1 25  4]
```

To do these same computations on a list, you would need to do something like a for loop, e.g. `[i**2 for i in x]` to square the elements. These vectorized computations are both much simpler to write and much faster to execute than for loops.

You can do math operations on two vectors at once, if they are the same shape. These operations are done element by element:

```
x = np.array([1,5,2])
y = np.array([2,3,1])

print(x - y)
print(x + y)
print(x/y)
print(x*2)
print(x**y)
```

```
[-1  2  1]
[3 8 3]
[0.5        1.66666667 2.        ]
[ 2 10  4]
[  1 125   2]
```

Numpy has equivalent functions to the base `math` library as well that can be applied to arrays, so `math.sqrt` has an equivalent `np.sqrt`, `math.log` has `np.log`, etc.:

```
x = np.array([1,5,2])

print(np.log(x))
print(np.log10(x))
print(np.sqrt(x))
print(np.exp(x))
```

```
[0.         1.60943791 0.69314718]
[0.      0.69897 0.30103]
[1.         2.23606798 1.41421356]
[  2.71828183 148.4131591    7.3890561 ]
```

Numpy arrays also have aggregation methods that come in handy, such as the sum, average, or standard deviation of the entire vector.

```
x = np.array([1,5,2])


print(x.mean())
print(x.std())
print(x.sum())
print(x.max())
print(x.min())
```

```
2.6666666666666665
1.699673171197595
8
5
1
```

There are more mathematical operations than this (such as `x.prod()`), you can check out `dir(x)` to see all of the possibilities.

> Note
>
> Many of those operations when you look at `dir(x)` are intended for matrix algebra operations, such as the trace (which is only defined for square matrices).

For these one dimensional arrays, they return a singular value. One method to keep in mind though is the `x.cumsum()` method, which returns a vector of the cumulative sums:

```
x = np.array([1,5,2])

print(x.cumsum())
```

```
[1 6 8]
```

There is also an `x.cumprod()` for the cumulative product of the elements.

Another convenient function commonly used in data analysis are quantiles. You can get the quantiles of a numpy array via the `np.quantile` function:

```
x = np.array([0,1,3,5])

# getting the 20th, 50th (median)
# and 80th quantile
print(np.quantile(x,[0.2,0.5,0.8]))
```

```
[0.6 2.  3.8]
```

Numpy arrays, like lists, are mutable. For a single dimensional array, you can access the items the same way as a list:

```
x = np.array([1,5,2,4])

print(x[0])
print(x[1])
print(x[-1])
print(x[1:])
print(x[:2])
print(x[1:3])
```

```
1
5
4
[5 2 4]
[1 5]
[5 2]
```

And because they are mutable, you can modify single elements of that array:

```
x = np.array([1,5,2,4])
print(x)
x[1] = -x[1]
print(x)
```

```
[1 5 2 4]
[ 1 -5  2  4]
```

One neat trick that you cannot do with lists, is that you can access numpy array indices repeatedly. So you can take a smaller list and generate repeated values:

```
x = np.array([1,5,2,4])
print(x[[1,1,0,0,0,3,2]])
```

```
[5 5 1 1 1 4 2]
```

You can also slice out specific elements of a list using boolean values:

```
x = np.array([1,5,2,4])
print(x[[True,False,True,False]])
```

```
[1 2]
```

This is often useful if you have some sort of boolean check on one array, and want to slice out values.

```
# slicing out values
# less than 3
x = np.array([1,5,2,4])
xs = x[x < 3]
print(xs)
```

```
[1 2]
```

If creating a sequence of numbers, the numpy functions of `arange` (similar to the `range` function in base python), `linspace`, `repeat`, and `tile` are useful to keep in mind. The function `np.linspace` generates equally spaced values between two endpoints. `np.repeat` generates replicates of the format `[1,1,...,2,2,...` and `np.tile` generates the form of `[1,2,1,2,...`.

```
print(np.arange(5))
print(np.arange(1,5))
print(np.linspace(1,5,4))

# for functions, can often pass
# a list OR a numpy array
print(np.repeat([1,2,3],3))

# can repeat different times
x = np.array([3,4,5])
print(np.repeat(x,[1,3,2]))

# if you want it different format
print(np.tile([1,2,3],3))
```

```
[0 1 2 3 4]
[1 2 3 4]
[1.         2.33333333 3.66666667 5.        ]
[1 1 1 2 2 2 3 3 3]
[3 4 4 4 5 5]
[1 2 3 1 2 3 1 2 3]
```

There is a repeat method for numpy arrays as well, so the `np.repeat(x,[1,3,2])` example you could also do `x.repeat([1,3,2])` and get the same result.

Numpy has several random data function generators that are often useful. You can generate random values between 0/1 or random coin flips. (It is a good idea to set the seed for reproducibility when doing simulations with randomized data.)

```
# setting seed for reproducibility
np.random.seed(10)

# random between (0,1)
print(np.random.rand(5)) # 5 observations

# random coin flip
```

```
print(np.random.binomial(1,0.5,size=(5,)))

# random integer value between 0-9
print(np.random.randint(10,size=(5,)))
```

```
[0.77132064 0.02075195 0.63364823 0.74880388 0.49850701]
[0 0 1 0 0]
[4 3 0 4 6]
```

For these examples, the `size` argument specifies the resulting number of random variates that are generated. Specifying the size argument as `(n,)` produces a vector of length `n`. I will show below working with matrices, so you could pass in `(rows,columns)` and get a different sized matrix for these random examples.

The `rand` function is a bit different than the others, in that it does not have any keyword arguments (you *only* pass the size of the resulting randomly generated numbers).

Another useful function is for randomly selecting items out of an array, `np.random.choice`:

```
# random choice out of a set
x = np.array(['a','b','c'])
print(np.random.choice(x,size=(5,)))

# default is with replacement, without
# can only do max 3 elements
print(np.random.choice(x,size=(3,),replace=False))
# this is like shuffling the vector

# can also specify unequal
# probabilities for choice
# very low prob for a
print(np.random.choice(x,size=(5,), p=[0.01,0.5,0.49]))
```

```
['a' 'c' 'c' 'b' 'a']
['b' 'c' 'a']
['c' 'c' 'b' 'b' 'c']
```

> Note
>
> Base python has a library, `random`, that has some similar random functions to those I show here for numpy. For example it has `random.random` (for random float values between 0 and 1) and `random.choice` (that is similar to the numpy choice function). These typically only generate *one*

random sample at a time, whereas numpy is vectorized, and can generate *many* random samples at once.

Here I show numpy arrays can have non-integer data contained in them. But, within an array they need to be all of the same type. Numpy on occasion will sometimes cause problems by implicitly converting objects to different types, such as integers to strings or integers to floats.

```
# showing implicit conversion
# converts entire array to strings
print(np.array([1,'a',2.4]))

# this converts ints to floats
print(np.array([1,2,3.1]))
```

```
['1' 'a' '2.4']
[1.  2.  3.1]
```

If you want to convert a float array to an integer array, you can use the `astype` method and specify the numeric type you want.

```
x = np.array([1,2,3.1])

# coerce to int
print(x.astype(int))

# floor, same result here
# but is still a float type
print(np.floor(x))
```

```
[1 2 3]
[1. 2. 3.]
```

So far I have *only* shown numpy arrays that are vectors, you can have matrices though with more dimensions. Here is a matrix with 3 rows and 2 columns:

```
# showing matrix, not just vector
x = np.array([[1,2],
              [3,4],
              [5,6]])

print(x)
print(x.shape) # rows,columns
```

```
[[1 2]
 [3 4]
 [5 6]]
(3, 2)
```

You can have higher than two dimensional arrays, but I do not show them here. To slice an array, you can use the usual notation for each dimension, but *seperate* slice notation for each dimension via a comma. Easier to show than to say!

```
x = np.array([[1,2],
              [3,4],
              [5,6]])

# select first row, all columns
print(x[0,:])

# select first column, all rows
print(x[:,0])

# select first two rows and first column
print(x[:2,0])
```

```
[1 2]
[1 3 5]
[1 3]
```

The same tricks for selecting out elements using booleans or indices work for multi-dimensional arrays as well.

Aggregate methods, like `.sum()` or `.mean()`, can be applied to the entire matrix, or *along* particular axes (on rows or columns in the two dimensional case):

```
x = np.array([[1,2],
              [3,4],
              [5,6]])

# mean for rows
print(x.mean(axis=1))

# mean for columns
print(x.mean(axis=0))

# sum for entire matrix
print(x.sum())
```

```
[1.5 3.5 5.5]
[3. 4.]
21
```

Note

You can *chain* together multiple aggregation commands. For example, if you did `x.mean(axis=0).max()`, this calculates the column means for the entire x array, and then identifies the maximum mean. This is something you can do with all objects in python. The approach `s = x.method1()` and `s.method2()` can be rewritten in one line using `x.method1().method2()`.

It works similar for quantiles as well. Here I show also another data operation task, called `ravel`, that turns a multi-dimensional array into a vector:

```
x = np.array([[1,2],
              [3,4],
              [5,6]])

# median for rows
med_rows = np.quantile(x,[0.5],axis=1)
print(med_rows)
print(med_rows.shape)

# if you want a single vector
# can use ravel
med_vec = med_rows.ravel()
print(med_vec)
```

```
[[1.5 3.5 5.5]]
(1, 3)
[1.5 3.5 5.5]
```

When working with multi-dimensional arrays, it can sometimes be useful to *broadcast* operations between different vectors. For example, say you wanted to mean detrend your data for each column. You can get the mean for each row, then subtract that value across each row in a vectorized manner.

```
x = np.array([[1,2],
              [3,4],
              [5,6]])

# mean for rows, keep the dims
xm = x.mean(axis=1,keepdims=True)
```

```
print(xm)
print(xm.shape)

# broadcasting operations across rows
print(x - xm)
```

```
[[1.5]
 [3.5]
 [5.5]]
(3, 1)
[[-0.5  0.5]
 [-0.5  0.5]
 [-0.5  0.5]]
```

You can do the same broadcasting for columns as well, here I median detrend the columns:

```
x = np.array([[1,2],
              [3,4],
              [5,6]])

# median for columns
xc = np.quantile(x,[0.5],axis=0)
print(xc)
print(xc.shape)

# broadcasting operations across columns
print(x - xc)
```

```
[[3. 4.]]
(1, 2)
[[-2. -2.]
 [ 0.  0.]
 [ 2.  2.]]
```

And finally, although not common for crime analysis (but good to know), you can additionally do matrix operations on the matrices, such as transposing the matrix, dot product for two matrices, the matrix norm, etc.

```
x = np.array([[1,2],
              [3,4],
              [5,6]])
```

```
# transpose
print(x.T)

# dot product
print(x.dot(x.T))

# matrix L2 norm on rows
print(np.linalg.norm(x,2,axis=1))

# can check out np.linalg
# for other matrix operations
# inverse, solve, etc.
```

```
[[1 3 5]
 [2 4 6]]
[[ 5 11 17]
 [11 25 39]
 [17 39 61]]
[2.23606798 5.         7.81024968]
```

For multidimensional arrays, `np.repeat` probably does not behave quite as you would expect, it unravels the two dimensional array to a vector. It may be the case that reshaping the vector will accomplish what you want.

```
x = np.array([[1,2],
              [3,4],
              [5,6]])

# repeat ravels array
np.repeat(x,2)

# can reshape to a different size
np.repeat(x,3).reshape(6,3)
```

```
array([[1, 1, 1],
       [2, 2, 2],
       [3, 3, 3],
       [4, 4, 4],
       [5, 5, 5],
       [6, 6, 6]])
```

If you want some type of repeated block structure, it may be using `np.concatenate` is what you need (passing in a *list* of numpy arrays). And also being familiar with Kronecker products is a useful trick to have up your sleeve.

```
x = np.array([[1,2],
              [3,4],
              [5,6]])


# concat rows
xr = np.concatenate([x,x])
print(xr)
print(xr.shape)

# concat columns
xc = np.concatenate([x,x],axis=1)
print(xc)
print(xc.shape)

# kronecker can do blocks
print(np.kron(x,[[1,1],[1,1]]))
```

```
[[1 2]
 [3 4]
 [5 6]
 [1 2]
 [3 4]
 [5 6]]
(6, 2)
[[1 2 1 2]
 [3 4 3 4]
 [5 6 5 6]]
(3, 4)
[[1 1 2 2]
 [1 1 2 2]
 [3 3 4 4]
 [3 3 4 4]
 [5 5 6 6]
 [5 5 6 6]]
```

It is best to deal with numpy arrays in a vectorized manner. But, if you need to do things like loop over an object, or turn two numpy arrays into a dictionary of key-value pairs, you can often treat numpy arrays the same as loops.

```
x = np.array([1,2,3])
y = np.array([4,5,6])

# can loop over array
for i in x:
    print(f'x value {i}')

# can turn into list if you
# need to
xl = x.tolist()
print(xl)

# can zip and turn into dict
di = dict(zip(x,y))
print(di)
```

```
x value 1
x value 2
x value 3
[1, 2, 3]
{1: 4, 2: 5, 3: 6}
```

In relations to sets, numpy has a function, `np.unique` that is convenient to return the unique values in an array. It also has the ability to return the *counts* of the number of times those values occur.

```
x = np.array([1,1,1,8,4,4])

# getting just the unique
# values
res = np.unique(x)
print(res)

# also returning the counts
res, cnt = np.unique(x, return_counts=True)
print(res, cnt)
```

```
[1 4 8]
[1 4 8] [3 2 1]
```

If you wanted to return this in a nicer table though, it is a bit more work. I have introduced `np.concatenate` earlier, but when working with vectors it is easier to use `np.stack`:

```
x = np.array([1,1,1,8,4,4])
res, cnt = np.unique(x, return_counts=True)

# this appends the two arrays together
# into one longer vector
# probably not what you want
table = np.concatenate([res,cnt])
print(table)

# to put these side by side
# into a matrix, use stack
table = np.stack([res,cnt],axis=1)
print(table)
```

```
[1 4 8 3 2 1]
[[1 3]
 [4 2]
 [8 1]]
```

If you wanted to put them stacked on top of each other, you could either just do `table.T` (to transpose the above table), or stack row vectors also using `np.stack([res,cnt],axis=0)`.

Even though I list all of these options for numpy arrays, I do not expect an individual to memorize them all. And there are many more than I show here – I am simply limiting the functions to showcase ones I have an occasional use for.

They are good to skim and be familiar with, in particular you should often strive to do operations on numpy arrays in a vectorized fashion. So if your code is looping to accomplish some mathematical task on vectors or matrices, it is good to review the code to see if it is possible to do that computation on the entire matrix at once.

6.2 pandas intro

The prior section on numpy arrays showed vector and matrix operations, mostly on numbers (integers and floats). When we are working with data, say a table of reported crimes, the days they are committed, and the geographic coordinates, we will have a mix of different data types.

In these scenarios, you probably want to work with the `pandas` python library. Pandas objects are called *dataframes*. Here is an example with a very simple table with both strings and numbers.

```
# import the pandas library
import pandas as pd

# creating simple data
```

```
x = [1,2,3]
y = ['a','b','c']

df = pd.DataFrame(zip(x,y),columns=['x','y'])
print(df)
```

```
   x  y
0  1  a
1  2  b
2  3  c
```

Similar to how most people do `import numpy as np`, it is common to abbreviate pandas as `pd`. Pandas dataframes have column names, here x and y, and row indices, here just a numeric index going from 0 to 2.

```
# checking out the dataframe contents

# columns
print(df.columns)

# row indices
print(df.index)

# variable types
print(df.dtypes)
```

```
Index(['x', 'y'], dtype='object')
RangeIndex(start=0, stop=3, step=1)
x     int64
y    object
dtype: object
```

Note that another way to get the dataframes columns is to do `list(df)` (in text, I will abbreviate a pandas dataframe as *df*). Pandas dataframes have many methods for exploring and manipulating data. A common first step when examining data is to look at the data via `df.head(k)` (to view the first k rows) or `df.tail(k)` to get the last k rows. By default, if you do `df.head()` or `df.tail()` it returns 5 rows

```
# first 2 rows
print(df.head(2))

# last 2 rows
```

```
print(df.tail(2))
```

```
   x  y
0  1  a
1  2  b
   x  y
1  2  b
2  3  c
```

You can access single columns of the dataframe by specifying `df['column_name']`.

```
# x column
print(df['x'])

# y column
print(df['y'])
```

```
0    1
1    2
2    3
Name: x, dtype: int64
0    a
1    b
2    c
Name: y, dtype: object
```

When you do `df['column_name']` this returns a pandas series object (similar to a numpy vector). You can operate on that series in a very similar fashion to how you would do numpy math operations. You often want to create *new* calculated columns, and you can add them back into pandas dataframes (dataframes are *mutable* objects).

```
# x column + 2
p2 = df['x'] + 2
print(p2)

# Can add this back into dataframe
df['x2'] = p2
print(df)
```

```
0    3
1    4
2    5
Name: x, dtype: int64
   x  y  x2
0  1  a   3
1  2  b   4
2  3  c   5
```

And you can do vectorized math operations on series against other series:

```
# dif to get back
df['xm'] = df['x'] - df['x2']

print(df)
```

```
   x  y  x2  xm
0  1  a   3  -2
1  2  b   4  -2
2  3  c   5  -2
```

Because dataframes are mutable, you can modify individual columns (overwrite them). Here I rewrite the `x2` column.

```
# overwrite the x2 column
df['x2'] = df['x']*df['x2']

print(df)
```

```
   x  y  x2  xm
0  1  a   3  -2
1  2  b   8  -2
2  3  c  15  -2
```

If you want to add in additional data outside of the dataframe, you can add in lists/tuples, or numpy arrays (as long as they are the correct number of rows to match up the dataframe). You can also add in a constant value.

```
# can add in lists/np arrays
df['le'] = ['ze','ye','xe']
df['ne'] = np.array([-3,-2,-1])
df['con'] = 1
```

```
print(df)
```

```
   x  y  x2  xm  le  ne  con
0  1  a   3  -2  ze  -3    1
1  2  b   8  -2  ye  -2    1
2  3  c  15  -2  xe  -1    1
```

Note that you can stuff whatever object you want into a list, and then put that into a column. For example, you could have a list of dictionaries.

```
# can have python objects in a list
# and add them to the dataframe
di = [{'d1': 'what', 'e1': 0},
      {'d2': 'yeah', 'e1': 1},
      {'d3': 'nope', 'e1': -1}]

df['di'] = di
print(df)
```

```
   x  y  x2  xm  le  ne  con                           di
0  1  a   3  -2  ze  -3    1   {'d1': 'what', 'e1': 0}
1  2  b   8  -2  ye  -2    1   {'d2': 'yeah', 'e1': 1}
2  3  c  15  -2  xe  -1    1  {'d3': 'nope', 'e1': -1}
```

The objects do not even need to be the same type (this is what allows you to have missing data fields for example).

```
# can have python objects in a list
# and add them to the dataframe
diff_types = [1,'a',set([1,2])]

df['dt'] = diff_types
print(df)
print(df.dtypes)
```

```
   x  y  x2  xm  le  ne  con                          di       dt
0  1  a   3  -2  ze  -3    1    {'d1': 'what', 'e1': 0}        1
1  2  b   8  -2  ye  -2    1    {'d2': 'yeah', 'e1': 1}        a
2  3  c  15  -2  xe  -1    1  {'d3': 'nope', 'e1': -1}  {1, 2}
x         int64
y        object
x2        int64
xm        int64
le       object
ne        int32
con       int64
di       object
dt       object
dtype: object
```

You can see it lists those dictionary and mixed type columns as `object` columns in pandas parlance.

Our dataframe is getting a bit crowded to print nicely, lets *drop* a few columns. Here I do a trick to drop all of the columns added *after* the first two `['x','y']`.

```
# getting the column names
# after x,y
ac = list(df)[2:]

# dropping columns
df.drop(columns=ac,inplace=True)
print(df)
```

```
   x  y
0  1  a
1  2  b
2  3  c
```

Here I have introduced an argument not previously seen `inplace=True`. So if I simply did `df.drop(columns=ac)`, this *will not* modify the `df` object, it will return a new dataframe. For certain dataframe methods, you can pass in the `inplace=True` argument to make it so the method modifies the dataframe in place. This forgoes the need to write something like `df = df.drop(columns=ac)`.

So far, we have only selected a single column, but you can select (and operate on) multiple columns at once. Instead of `df['x']`, you can pass in a list of column names, e.g. `df[['x','y']]`.

```
# adding in a second numeric column
df['n'] = [3,5,8]
```

```
# getting two numeric columns
tc = df[['x','n']]
print(tc)

# can manipulate and add these back
# in all at once
df[['x2','n2']] = tc*2
print(df)
```

```
   x  n
0  1  3
1  2  5
2  3  8
   x  y  n  x2  n2
0  1  a  3   2   6
1  2  b  5   4  10
2  3  c  8   6  16
```

Note that if you did `df[list(tc)] = tc`, it would have overwritten the initial `x` and `n` columns in the `df` dataframe, instead of creating two new columns.

> Note
>
> On occasion, you want to work with a single column, but still want the returned object to be a dataframe. You can do `df[['column_name']]` – e.g. pass a list with a single column name. This will return a dataframe object instead of a pandas series.

6.3 pandas filtering

Filtering is what I use to describe *selecting out specific rows* of a dataframe. Here to illustrate, I am going to add in a few pieces to our example dataframe, in particular instead of the default integer index, e.g. `[0,1,2]`, I am going to set a specific index of `['A','B','C']` on the rows.

```
# new dataframe, using index as letters
x = [1,2,3]
y = ['z','y','x']
z = ['1/1/2023','1/2/2023','1/3/2023']
indexVals = ['A','B','C']
df = pd.DataFrame(zip(x,y,z),
                  columns=['x','y','z'],
                  index=indexVals)

# note that the z field is a string currently
```

```
print(df)
```

```
   x  y         z
A  1  z  1/1/2023
B  2  y  1/2/2023
C  3  x  1/3/2023
```

To select out specific rows, there are three general approaches. The simplest approach is to select based on an array of booleans:

```
# select based on array
bool = [True,False,True]
print(df[bool]) # selects 1st and 3rd rows
```

```
   x  y         z
A  1  z  1/1/2023
C  3  x  1/3/2023
```

This is most commonly used when doing a boolean operation on a pandas series:

```
# select based pandas series condition
bool = df['x'] <= 2
print(df[bool]) # selects 1st and 2nd rows
```

```
   x  y         z
A  1  z  1/1/2023
B  2  y  1/2/2023
```

This is convenient to chain together multiple conditions. Boolean operations on pandas objects are the same as for base python objects, they just operate on each row one at a time.

```
# multiple conditions
b1 = (df['x'] <= 2) & (df['y'] == 'z')
print(df[b1]) # selects just the first row

# or condition
print('\nSelecting First two rows')
b2 = (df['x'] <= 1) | (df['y'].isin(['z','y']))
print(df[b2]) # selects first two rows
```

```
   x  y          z
A  1  z  1/1/2023

Selecting First two rows
   x  y          z
A  1  z  1/1/2023
B  2  y  1/2/2023
```

I have introduced a new operation here that I use often, `df['x'].insin(my_list)`. If you have many equality checks, it is easier to use `isin` than it is to check equality for multiple objects.

> Note
>
> To reverse a pandas (or numpy) series of booleans, you can use `~bool`. So for example, *is not in* would be `~df['y'].isin(['z','y'])`.

Sometimes this can be confusing, because `df['string']` returns a column whereas `df[bool]` filters rows and returns all the columns. In case you want to select out certain columns and rows at the same time, you can use `df.iloc` or `df.loc`. To keep it simple the difference between the two, `df.iloc` works with either booleans or integers, and `df.loc` works with strings.

Here is an example with `df.iloc` on our prior dataframe. The parameters are `df.iloc[rows,columns]` where rows and columns are either integers or booleans.

```
# selecting out rows 1 and 3, columns 2 and 1
# remember 0 based indexing!
print(df.iloc[[0,2],[2,0]])

# can also be booleans
print('') # spaces in-between outputs!
print(df.iloc[[True,False,True],[True,False,True]])

# can have a mix
print('')
print(df.iloc[[True,False,True],[2,0]])
```

```
          z  x
A  1/1/2023  1
C  1/3/2023  3

   x         z
A  1  1/1/2023
C  3  1/3/2023

          z  x
A  1/1/2023  1
C  1/3/2023  3
```

To select out *all* either rows or columns, you can use a `:`, also can use the same slice notation for lists.

```
# first two rows, all columns
print(df.iloc[:2,:])

# all rows, second column until end
print('')
print(df.iloc[:,1:])

# last 2 rows, last 2 columns
print('')
print(df.iloc[-2:,-2:])
```

```
   x  y         z
A  1  z  1/1/2023
B  2  y  1/2/2023

   y         z
A  z  1/1/2023
B  y  1/2/2023
C  x  1/3/2023

   y         z
B  y  1/2/2023
C  x  1/3/2023
```

Note with these selections for booleans, it works the same way as passing a list vs passing a numpy vector or pandas series. E.g. `df.iloc[bool,[0,2]]` works the same if `bool` is a list of booleans or a pandas series.

The method `df.loc` is similar, but operates using strings. Since I have made the index for the current dataframe `['A','B','C']`

```
# selecting base on index
print(df.loc[['A','C'],:])

# ditto for columns
print('')
print(df.loc[['C','B'],['z','x']])

# can use boolean
print('')
print(df.loc[df['x'] <= 2,['y','z']])
```

```
   x  y         z
A  1  z  1/1/2023
C  3  x  1/3/2023

          z  x
C  1/3/2023  3
B  1/2/2023  2

   y         z
A  z  1/1/2023
B  y  1/2/2023
```

Because dataframes often have a default index that is `[0,1,....,n]` (a range index), row selection between loc and iloc will be the same in those two scenarios. But column selection will not be the same.

> Note
>
> When you create a new dataframe from an old one, such as `dfnew = df_old[selection]`, when you subsequently do operations on `dfnew` it is common to get a warning in pandas along the line of `SettingWithCopyWarning`. To prevent this warning, you often want to do `dfnew = df_old[selection].copy()` – this makes it so `dfnew` is a totally seperate object than `df_old`, so there is no potential issues with mutating one object and it propagating to the other.

If you want to change certain objects in a dataframe, you often want to use `loc` or `iloc` and set those values directly:

```
# reseting the df dataframe
x = [1,2,3]
y = ['z','y','x']
z = ['1/1/2023','1/2/2023','1/3/2023']
indexVals = ['A','B','C']
df = pd.DataFrame(zip(x,y,z),
```

```
                        columns=['x','y','z'],
                        index=indexVals)

# can set items directly
df.iloc[[0,2],1] = ['t','u']
print(df)
```

```
   x  y         z
A  1  t  1/1/2023
B  2  y  1/2/2023
C  3  u  1/3/2023
```

One gotcha I want to point out on this section, is that when you have dataframes with indexes, they can sometimes not behave as you would expect. So here is a mistake I sometimes make:

```
# dataframe with index 2,3,4
x = [1,2,3]
y = ['z','y','x']
df = pd.DataFrame(zip(x,y),
                  columns=['x','y'],
                  index=[2,3,4])

# creating a new series
new_vals = pd.Series([4,5,6])

# This is probably not what you want
df['n'] = new_vals
print(df)
```

```
   x  y    n
2  1  z  6.0
3  2  y  NaN
4  3  x  NaN
```

Whoa – what happened here? `df['n']` has values `6.0, nan, nan`. This is because when assigning values between two pandas objects, it implicitly matches on the indices for the two objects. So here `new_values` has the default index `[0,1,2]`, whereas `df` has indices `[2,3,4]`.

There are multiple ways to solve this, you could *reset* the original dataframe, so its indices are now `[0,1,2]`:

```
# dataframe with index 2,3,4
x = [1,2,3]
```

```
y = ['z','y','x']
df = pd.DataFrame(zip(x,y),
                  columns=['x','y'],
                  index=[2,3,4])

# creating a new series
new_vals = pd.Series([4,5,6])

# here I reset the original dataframe
df.reset_index(inplace=True)
df['n'] = new_vals
print(df)
```

```
   index  x  y  n
0      2  1  z  4
1      3  2  y  5
2      4  3  x  6
```

If you don't want the `index` column, `df.reset_index(inplace=True,drop=True)` will prevent that column from being added to the resulting dataframe. Another solution would be to assign based on the list of items, not the pandas series object:

```
# dataframe with index 2,3,4
x = [1,2,3]
y = ['z','y','x']
df = pd.DataFrame(zip(x,y),
                  columns=['x','y'],
                  index=[2,3,4])

# creating a new series
new_vals = pd.Series([4,5,6])

# here I assign based on list
df['n'] = new_vals.tolist()
print(df)
```

```
   x  y  n
2  1  z  4
3  2  y  5
4  3  x  6
```

And a final example is you could set the index of the pandas series object directly as well.

```
# dataframe with index 2,3,4
x = [1,2,3]
y = ['z','y','x']
df = pd.DataFrame(zip(x,y),
                  columns=['x','y'],
                  index=[2,3,4])

# creating a new series with index
new_vals = pd.Series([4,5,6], index=df.index)

# here I assign based on list
df['n'] = new_vals.tolist()
print(df)
```

```
   x  y  n
2  1  z  4
3  2  y  5
4  3  x  6
```

This error is a common gotcha though (at least for me) when working with pandas dataframes that have non-default range indexes and series.

The final example I want to describe in this section is *sorting* a dataframe. This is convenient in data exploration. I commonly do something like this to check out the top-k rows in a dataset.

```
# dataframe with index 2,3,4
x = [1,2,3]
y = ['z','y','x']
indexVals = ['A','B','C']
df = pd.DataFrame(zip(x,y),
                  columns=['x','y'],
                  index=indexVals)

# sorting inplace
df.sort_values(by='y',inplace=True)
print(df)
```

```
   x  y
C  3  x
B  2  y
A  1  z
```

This is a simplified example to help teach you how the operation works, but say you had a dataframe that had street segments and counts of crime. You could sort the dataframe, `df.sort_values(by='Crime',ascending=False,inplace=True)` then look at the top 10 crime street segments in your jurisdiction via `df.head(10)`.

If you want to reset the index, you can use `df.sort_values(by='x',ignore_index=True)`. And if you want to sort by multiple fields, you can pass in lists to the `by` argument and `ascending` arguments. So say you want to sort by violent crimes and then by property crimes, you could do below for example.

```
df.sort_values(by=['Violent','Property'],ascending=False)
```

6.4 pandas field functions

Basic mathematic operations on numeric fields in pandas are the same as they are for numpy vectors (they are numpy vectors under the hood).

```
# pandas series instead of numpy array
x = pd.Series([1,5,2])

print(x - 1)
print(x + 1)
print(x/2)
print(x*2)
print(x**2)
```

```
0    0
1    4
2    1
dtype: int64
0    2
1    6
2    3
dtype: int64
0    0.5
1    2.5
2    1.0
dtype: float64
0     2
1    10
2     4
dtype: int64
0     1
1    25
2     4
dtype: int64
```

Note again a common gotcha with pandas series (same as I showed with dataframes prior), they need to line up in their indices, or you get unexpected results. Here it results in a new vector with 6 rows all missing data.

```
# without index statement, default is 0,1...n
x = pd.Series([1,2,3],index=['a','b','c'])
y = pd.Series([4,5,6],index=[1,2,3])

# will be all missing, since indices don't line up
print(x + y)
```

```
a    NaN
b    NaN
c    NaN
1    NaN
2    NaN
3    NaN
dtype: float64
```

There are *many* different helper functions though (in addition to the same mathematical operations) that pandas has. I will go through some of the common ones I use.

One is turning a categorical variable into a series of dummy variables (equal to 1 when present in a row, vs 0 when not present). This is convenient for aggregation and different regression modelling tasks.

```
x = ['a','b','c']
df = pd.DataFrame(x,columns=['x'])
dum_x = pd.get_dummies(df['x'])

# can add this entire matrix back in
df[list(dum_x)] = 1*dum_x # converting to 0/1's
print(df)
```

```
   x  a  b  c
0  a  1  0  0
1  b  0  1  0
2  c  0  0  1
```

The function `pd.get_dummies` has many other options worth investigating, such as dropping rare categories, encoding missing data, and how the fields are named.

> Note
>
> The function `pd.get_dummies(df['x'])` returns a dataframe of boolean True/False results. Sometimes it is more convenient to work with 0/1 data (such as when printing the results). In that case, you can do `1*bool` (where `bool` can be a dataframe or a series), like I do above, to get the 0/1 values.

Pandas has the ability to deal with missing data. Missing data in pandas dataframes can be encoded as `np.nan` (numpy's not a number), or `None`. To replace missing values with a common value across the entire pandas series, you can use `x.fillna(value)`.

```
x = [1,2,3]
y = [1,np.nan,3]
df = pd.DataFrame(zip(x,y),columns=['x','y'])
df['z'] = df['x'] + df['y']

print('Will have missing for z 2nd row')
print(df)

print('\nFilling in missing')
df['z'] = df['z'].fillna(-1)
print(df)
```

```
Will have missing for z 2nd row
   x    y    z
0  1  1.0  2.0
1  2  NaN  NaN
2  3  3.0  6.0

Filling in missing
   x    y    z
0  1  1.0  2.0
1  2  NaN -1.0
2  3  3.0  6.0
```

There is a function that works on entire dataframes as well – for example here you could have done `df.fillna(-1,inplace=True)`. If you only want to fill in missing for a single column though, you need to assign it back into the dataframe as I show here.

Since you can pass in any value to the `fillna` method, to do something like replace missing with the median would be as simple as `df['z'].fillna(df['z'].median())`.

Other missing value functions to be familiar with are forward fill `.ffill`, back-fill `.bfill`, and interpolate `.interpolate`. Those are common missing data imputation techniques for data that is time ordered, so values are filled in based on values prior or forward in time (or in between).

The next field based operation I am going to show is replacing values with a dictionary. Here I add in different labels to the dataframe.

```
x = ['a','b','c']
df = pd.DataFrame(x,columns=['x'])

# replacing x with labels
rep_di = {'a': 'LabelA',
          'b': 'LabelB'}

df['lab'] = df['x'].replace(rep_di)

# Note that c is not replaced
print(df)
```

```
   x     lab
0  a  LabelA
1  b  LabelB
2  c       c
```

The keys of the dictionary should map directly to the objects in the pandas dataframe. They can be numeric values as well, but most of the time people use this with categorical/string variables.

Sometimes instead of replacing with a label, you want to bin a numeric field into different categories. The `cut` and `qcut` methods to accomplish that task are convenient.

```
np.random.seed(10) # setting random seed
x = np.arange(10) # integers 0 to 9
y = np.random.rand(10) # random uniform
df = pd.DataFrame(zip(x,y),columns=['x','y'])

# binning into categories
df['binx'] = pd.cut(df['x'],[-0.1,3,7,10])

# quantile cut, 3 equal bins
df['biny'] = pd.qcut(df['y'],3)

print(df)
```

```
   x         y          binx                              biny
0  0  0.771321  (-0.1, 3.0]                       (0.634, 0.771]
1  1  0.020752  (-0.1, 3.0]  (0.019799999999999998, 0.198]
2  2  0.633648  (-0.1, 3.0]                       (0.198, 0.634]
3  3  0.748804  (-0.1, 3.0]                       (0.634, 0.771]
4  4  0.498507   (3.0, 7.0]                       (0.198, 0.634]
5  5  0.224797   (3.0, 7.0]                       (0.198, 0.634]
6  6  0.198063   (3.0, 7.0]  (0.019799999999999998, 0.198]
7  7  0.760531   (3.0, 7.0]                       (0.634, 0.771]
8  8  0.169111  (7.0, 10.0]  (0.019799999999999998, 0.198]
9  9  0.088340  (7.0, 10.0]  (0.019799999999999998, 0.198]
```

The behavior for cut is `(,]`, that is the left end of the interval is open, and the right interval is closed. If I had done `pd.cut(df['x'],[0,3,7,10])`, the first row would not be assigned, because it is "greater than 0". Here I solve this by using a value slightly below the minimum, but there are different options for `pd.cut` to solve this. I encourage the users to either do `help(pd.cut)`, or google `pandas cut` to see all of the options for the function.

The function `pd.qcut` cuts up a quantitative value according to the quantiles the user specifies or the number of bins. Here since I specify 3 bins, you will have one bin with four rows, and the other two bins with three rows apiece. (Note with data with potential ties, you need to be careful with this function, as it could fail as the quantiles are duplicated. One potential solution in that scenario is to add a small random value to break the ties.)

The next example I want to give with pandas functions are special methods for dealing with strings. If you have a column of strings, you can do `df['x'].str.method`, where many of the same methods available for strings in basic python objects you can use row-wise on pandas series.

```
x = ['a12 ','bbbbb',' c']
y = ['1','2','3']
df = pd.DataFrame(zip(x,y),columns=['x','y'])

# string methods, such as len, trim
print(df['x'].str.len())
print(df['x'].str.strip())  # gets rid of whitespace on ends
print(df['y'].str.zfill(3)) # adds leading zeros
```

```
0    4
1    5
2    2
Name: x, dtype: int64
0      a12
1    bbbbb
2        c
Name: x, dtype: object
0    001
1    002
2    003
Name: y, dtype: object
```

Note that because you can chain methods together, you can add together multiple string methods, but you have to call `.str` each time, such as `df['x'].str.strip().str.len()` to strip whitespace and then get the string length without whitespace.

If you just do `df['x'].str`, this is like working with the string object directly, so you can do slice notation on the strings in the field:

```
x = ['a12 ','bbbbb',' c']
y = ['1','2','3']
df = pd.DataFrame(zip(x,y),columns=['x','y'])

# can slice the string directly
print(df['x'].str[:2]) # first 2 chars
print(df['x'].str[-2:]) # last 2
print(df['x'].str[1]) # 2nd char
```

```
0    a1
1    bb
2     c
Name: x, dtype: object
0    2
1    bb
2     c
Name: x, dtype: object
0    1
1    b
2    c
Name: x, dtype: object
```

You can concatenate fields together.

```
x = ['a12 ','bbbbb',' c']
y = ['1','2','3']
df = pd.DataFrame(zip(x,y),columns=['x','y'])

# can concatenate strings together
df['z'] = df['x'] + "_" + df['y']
print(df)
```

```
       x  y        z
0   a12   1   a12 _1
1  bbbbb  2  bbbbb_2
2      c  3      c_3
```

You can convert other values to strings using `.astype(str)`. But a more common procedure is to want to convert a string that is a set of numbers to an actual numeric field. For that you can use `pd.to_numeric`:

```
x = ['a12 ','bbbbb',' c']
y = ['1','2','3'] # this is a string field
z = [1,2,3] # this is a numeric field
df = pd.DataFrame(zip(x,y,z),columns=['x','y','z'])

# can convert number to string using astype
print(df['z'].astype(str))

# can convert string to number using function
df['zn'] = pd.to_numeric(df['y'])/2
print(df)
```

```
0    1
1    2
2    3
Name: z, dtype: object
       x  y  z   zn
0   a12   1  1  0.5
1  bbbbb  2  2  1.0
2      c  3  3  1.5
```

Here you could do `df['y'].astype(int)` as well, but `pd.to_numeric` has some nice features to handle conversions (such as values that need to be set to missing), that using the `astype` conversion may cause an error.

The final field manipulation example I want to give is working with time data in pandas. You can convert string data that are dates (or datetime stamps) to a date/datetime field in pandas.

```
x = ['1/1/2023','1/2/2023','1/3/2023'] # just dates
y = ['2023-01-01 15:00',
     '2023-01-03 12:00',
     '2023-01-06 23:59'] # dates and times
df = pd.DataFrame(zip(x,y),columns=['x','y'])

# at first these are string fields, but can convert to dates
df['x'] = pd.to_datetime(df['x'])
df['y'] = pd.to_datetime(df['y'])
print(df)
```

```
           x                   y
0 2023-01-01 2023-01-01 15:00:00
1 2023-01-02 2023-01-03 12:00:00
2 2023-01-03 2023-01-06 23:59:00
```

To access certain date objects, similar to strings you use `df['x'].dt` to access the underlying objects.

```
# using prior df
print(df['x'].dt.year)
print(df['x'].dt.month)
print(df['y'].dt.hour)
print(df['y'].dt.minute)
```

```
0    2023
1    2023
2    2023
Name: x, dtype: int64
0    1
1    1
2    1
Name: x, dtype: int64
0    15
1    12
2    23
Name: y, dtype: int64
0     0
1     0
2    59
Name: y, dtype: int64
```

Crime analysis typically does not need to worry about down to the seconds, but depending on the source of the data, the database may record the fields to that precision (such as CAD data when a 911 call occurs).

You can add or substract two datetime fields – here `df['x']` is just treated as having hours and minutes at `00:00`:

```
# can calculate diff
time_diff = df['y'] - df['x']
print(time_diff)

# This has objects, such as days and seconds
print(time_diff.dt.days)
print(time_diff.dt.total_seconds())
```

```
0   0 days 15:00:00
1   1 days 12:00:00
2   3 days 23:59:00
dtype: timedelta64[ns]
0    0
1    1
2    3
dtype: int64
0     54000.0
1    129600.0
2    345540.0
dtype: float64
```

Sometimes you want to add a fixed number to a particular date, e.g. get the date one week from the current date. You can use `pd.Timedelta` to do that, this example adds in a fixed number of days.

```
# adding in 7 days to each, unit can be different types
# of units, like seconds, months, etc.
df['z'] = df['x'] + pd.Timedelta(7,unit='D')

print(df[['x','z']])
```

```
           x          z
0 2023-01-01 2023-01-08
1 2023-01-02 2023-01-09
2 2023-01-03 2023-01-10
```

And this example adds a variable number of hours depending on a particular row value.

```
# add in variable number of hours
df['num_hours'] = [3,-1,2]
df['yv'] = df['y'] + pd.to_timedelta(df['num_hours'],unit='H')

# This has objects, such as days and seconds
print(df[['y','num_hours','yv']])
```

```
                    y  num_hours                  yv
0 2023-01-01 15:00:00          3 2023-01-01 18:00:00
1 2023-01-03 12:00:00         -1 2023-01-03 11:00:00
2 2023-01-06 23:59:00          2 2023-01-07 01:59:00
```

Note

Sometimes it is convenient to make temporal data with regular intervals. You can construct every day between two dates via `pd.date_range('1/1/2023','1/10/2023')`. You can construct weeks via changing the frequency argument `pd.date_range('1/1/2023','1/14/2023',freq='7D')`, or months via `pd.date_range('1/1/2022','1/1/2023',freq='1M')`. Similar to `np.arange`, you need to be careful how the end of the range of numbers is generated based on the particular argument functions.

6.5 pandas aggregations

So far the field operations were vectorized – they applied operations on each row independently for a pandas dataframe. Each function took in say 10 rows of data, and produced a new series of 10 values.

The operations I am going to show in this section produce *aggregations*; they take in a set of data and produce a smaller number of rows that are some type of summary of the data.

The prior statistical aggregations I showed in the numpy section also work for pandas series. So if you do `df['x'].mean()`, this will produce the mean of a particular column. But often we want more complicated summaries via pandas, in particular dealing with aggregations by different categorical variables or multiple variables.

The first aggregation I often use in exploratory data analysis for categorical columns is the method `.value_counts()`

```
x = ['a']*3 + ['b','c']
y = [1,2,3,4,5]
df = pd.DataFrame(zip(x,y),columns=['x','y'])

# value counts
```

```
print(df['x'].value_counts())
```

```
a    3
b    1
c    1
Name: x, dtype: int64
```

This produces a pandas *series*, where the *indices* are the original values, and the values within the series are counts of unique items. I often don't use this in further data processing, as the conversion to an index can sometimes change the objects (if you want to get the unique values in a pandas series, you can do `pd.unique(df['x'])`.

The most common workhorse for aggregating data in pandas dataframes is the `.groupby` method. This aggregates a variable and performs some type of statistical summary across the groups.

```
x = ['a']*3 + ['b','c']
y = [1,2,3,4,5]
z = [10,11,12,13,14]
df = pd.DataFrame(zip(x,y,z),columns=['x','y','z'])

# groupby and mean
gdf = df.groupby('x').mean()
print(gdf)
```

```
     y     z
x
a  2.0  11.0
b  4.0  13.0
c  5.0  14.0
```

Note that this generates the means for *both* the variables `y` and `z`. It is not uncommon to only want a specific variable. Here I also show an option to not generate the aggregate dataframe with the index set to the grouping variable.

```
x = ['a']*3 + ['b','c']
y = [1,2,3,4,5]
z = [10,11,12,13,14]
df = pd.DataFrame(zip(x,y,z),columns=['x','y','z'])

# only selecting y, x will be a variable as well
# aggregating to the sum
gdf = df.groupby('x',as_index=False)['y'].sum()
print(gdf)
```

```
   x  y
0  a  6
1  b  4
2  c  5
```

You can aggregate by multiple variables, instead of just one, by passing in a list of variables to `.groupby`.

```
x = ['a']*3 + ['b','b']
y = [1,2,3,4,5]
z = [10,10,11,11,11]
df = pd.DataFrame(zip(x,y,z),columns=['x','y','z'])

# can groupby multiple columns
gdf = df.groupby(['x','z'],as_index=False)['y'].min()
print(gdf)
```

```
   x   z  y
0  a  10  1
1  a  11  3
2  b  11  4
```

If you did not specify `as_index=False` in this example, it would return a *multi-index*. I don't have much use for those in practice, so I personally more often than not use the `as_index=False` option for groupby.

You can also have the function return multiple aggregate statistics. Here is an example returning the mean, the sum, and the total number of observations (via the `size` function).

```
x = ['a']*3 + ['b','b']
y = [1,2,3,4,5]
z = [10,10,11,11,11]
df = pd.DataFrame(zip(x,y,z),columns=['x','y','z'])

# will return multiple aggregation summaries
gdf = (df.groupby('x',as_index=False)['y']
          .agg(["sum","size","mean"]))
print(gdf)
```

```
   sum  size  mean
x
a    6     3   2.0
b    9     2   4.5
```

You can see I did a trick here as well to split the dataframe processing across multiple lines. If you wrap the operation in a parenthesis (similar to the long string trick I have previously shown), you can split very long operations onto multiple lines.

If you do summaries on multiple variables, you can see that you will have a nested structure now for the column names.

```
x = ['a']*3 + ['b','b']
g = [1,1,2,2,2]
y = [1,2,3,4,5]
z = [10,10,11,11,11]
df = pd.DataFrame(zip(x,y,z,g),columns=['x','y','z','g'])

# does agg for both y and z (if you have strings, will fail!)
gdf = (df.groupby(['x','g'],as_index=False)
         .agg(["sum","size","mean"]))
print(gdf)
```

```
      y              z
    sum size mean sum size  mean
x g
a 1   3    2  1.5  20    2  10.0
  2   3    1  3.0  11    1  11.0
b 2   9    2  4.5  22    2  11.0
```

To select a specific column in this aggregated table, you would need to do something like `gdf['y']['sum']` – so first select the `y` variable, and then select the sub-variable name `sum`.

You can additionally create your own functions, and pass that into `.agg`. Here I make a range function, as well as min/max.

```
def rangev(x):
  return max(x) - min(x)

agg_funcs = {'range': rangev,
             'min': min,
             'max': max}

x = ['a']*3 + ['b','b']
g = [1,1,2,2,2]
y = [1,2,3,4,5]
z = [10,10,11,11,11]
df = pd.DataFrame(zip(x,y,z,g),columns=['x','y','z','g'])

# can do your own functions for aggregation
```

```
gdf = df.groupby('x',as_index=False)['y'].agg(agg_funcs)
print(gdf)
```

```
   x  range  min  max
0  a      2    1    3
1  b      1    4    5
```

This format only works when aggregating a single variable. You can pass in a dictionary to specify specifically *what* variables to do particular aggregations on (and their functions),

```
def rangev(x):
  return max(x) - min(x)

# keyword is final name
# values are (field,aggfunc)
agg_var = {'ymin': ('y',min),
           'ymax': ('y',max),
           'zrange': ('z',rangev)}

x = ['a']*3 + ['b','b']
g = [1,1,2,2,2]
y = [1,2,3,4,5]
z = [10,10,11,11,11]
df = pd.DataFrame(zip(x,y,z,g),columns=['x','y','z','g'])

# if agg func multiple, needs to look a bit different
# if aggregating more than one variable, should pass in
# as list
gdf = df.groupby('x',as_index=False)[['y','z']].agg(**agg_var)
print(gdf)
```

```
   x  ymin  ymax  zrange
0  a     1     3       1
1  b     4     5       0
```

Sometimes you want to add in the aggregated data into the original dataframe. Such as the mean or the sum per group. You can also do that via groupby, but by using the `transform` method instead of `agg`:

```
def median(x):
    return x.median()

x = ['a']*3 + ['b','b']
```

```
y = [1,2,3,4,5]
df = pd.DataFrame(zip(x,y),columns=['x','y'])

# can add in grouped stats back to original dataframe
df['medy_byx'] = (df.groupby('x',as_index=False)['y']
                   .transform(median))
print(df)
```

```
   x  y  medy_byx
0  a  1       2.0
1  a  2       2.0
2  a  3       2.0
3  b  4       4.5
4  b  5       4.5
```

Another common data manipulation is *ranking* per group. This is convenient for say identifying the ranking of charges per a crime incident (with the top charge being the highest ranked charge category). Pandas has methods to do this.

```
x = ['a']*3 + ['b','b']
y = [1,3,0,2,1]
df = pd.DataFrame(zip(x,y),columns=['x','y'])

# can do ranks within groups
df['ranky_byx'] = df.groupby('x',as_index=False)['y'].rank()
print(df)
```

```
   x  y  ranky_byx
0  a  1        2.0
1  a  3        3.0
2  a  0        1.0
3  b  2        2.0
4  b  1        1.0
```

If you check out the pandas rank command (you could do `help(df['y'].rank)`, but I think the web document when just googling `pandas rank` and going to the official online documents is easier to peruse), you can see options such as ranking in ascending/descending, and how to treat ties.

Two last types of aggregations I want to show are *dropping duplicates* and *crosstabs*. Dropping duplicates you can either specify to drop based on exact same values across the entire dataframe (the default), or drop duplicates based on one or more specific columns.

```
x = ['a']*3 + ['b','b']
y = [1,3,3,2,2]
df = pd.DataFrame(zip(x,y),columns=['x','y'])

# dropping duplicates for entire dataframe
dedup1 = df.drop_duplicates()
print(dedup1) # two rows dropped

print('\nDropping dups based on var subset')
df.drop_duplicates(subset='x',inplace=True)
print(df) # should only have two rows
```

```
   x  y
0  a  1
1  a  3
3  b  2

Dropping dups based on var subset
   x  y
0  a  1
3  b  2
```

If you drop duplicates based on specific columns, you need to keep in mind what other variables are selected. You can either keep the first row or the last row. As opposed to aggregating, this is another way to select a top charge – e.g. if you sort the dataframe at first, and then drop duplicates within the incident ID.

The final example I want to show is crosstabs. This is common to look at the frequency of two categorical variables.

```
x = ['a']*3 + ['b']*3 + ['c']*2
y = [1,1,1,1,1,2,1,1,2]
df = pd.DataFrame(zip(x,y),columns=['x','y'])
ct = pd.crosstab(df['x'],df['y'])
print(ct)
```

```
y  1  2
x
a  3  0
b  2  1
c  2  0
```

It is also possible to do `df.groupby(['x','y']).size()` in this example. But that will return the data in a different format, as well as not have pairs of groups that have zero values.

6.6 pandas merging and reshaping

Sometimes you work with not just a single dataframe, but multiple dataframes. You may have one dataframe that is crime incident data, and you want to join the information from the officer who responded to the crime incident dataset. You would use the `merge` function to accomplish that (pretend `g` in this example is an officer id):

```
# first dataframe
x = [1,2,3]
y = ['a','b','c']
g = [1,1,2]
df = pd.DataFrame(zip(x,y,g),columns=['x','y','g'])

# second dataframe
v2 = [10,11]
g2 = [1,2]
df2 = pd.DataFrame(zip(v2,g2),columns=['v2','g'])

# now merging the two together
dfm = df.merge(df2,on='g')
print(dfm)
```

```
   x  y  g  v2
0  1  a  1  10
1  2  b  1  10
2  3  c  2  11
```

This particular merge you can see `df2` only has two rows, and it merges one to many to the original `df` dataframe. Equivalently here you could do `pd.merge(df,df2,on='g')` and it would produce the same result.

Another type of merge is an inner merge – the resulting dataframe *only* contains rows that were available in each dataframe.

```
# first dataframe
x = [1,2,3]
y = ['a','b','c']
g = [1,2,3]
df = pd.DataFrame(zip(x,y,g),columns=['x','y','g'])

# second dataframe only has 1/2
v2 = [10,11]
g2 = [1,2]
df2 = pd.DataFrame(zip(v2,g2),columns=['v2','g'])
```

```
# this does not contain row 3 now from df
dfm = df.merge(df2,on='g',how='inner')
print(dfm)
```

```
   x  y  g  v2
0  1  a  1  10
1  2  b  2  11
```

You may however wish to keep those rows, and just have it as missing data. To do that, you can pass in `how='left'`, and also include an indicator column that provides information on the type of merge.

```
# first dataframe
x = [1,2,3]
y = ['a','b','c']
g = [1,2,3]
df = pd.DataFrame(zip(x,y,g),columns=['x','y','g'])

# second dataframe only has 1/2
v2 = [10,11]
g2 = [1,2]
df2 = pd.DataFrame(zip(v2,g2),columns=['v2','g'])

# this contains missing data for row 3
dfm = df.merge(df2,on='g',how='left',
      indicator='merge_info')
print(dfm)
```

```
   x  y  g    v2 merge_info
0  1  a  1  10.0       both
1  2  b  2  11.0       both
2  3  c  3   NaN  left_only
```

You can merge on multiple columns, by passing in a list of column names to the `on` kwarg:

```
# first dataframe
x = [1,2,3]
y = ['a','b','c']
g = [1,2,3]
h = ['Z','Y','Y']
df = pd.DataFrame(zip(x,y,g,h),
                  columns=['x','y','g','h'])
```

```
# second dataframe only has 1/2
v2 = [10,11]
g2 = [1,2]
h2 = ['Z','Z']
df2 = pd.DataFrame(zip(v2,g2,h2),columns=['v2','g','h'])

# only matches on row1 now
dfm = df.merge(df2,on=['g','h'],how='left',
        indicator='merge_info')
print(dfm)
```

```
   x  y  g  h    v2 merge_info
0  1  a  1  Z  10.0       both
1  2  b  2  Y   NaN  left_only
2  3  c  3  Y   NaN  left_only
```

If you have field names that are the same between dataframes, but *are not* merge keys, pandas will rename them with different suffixes to prevent name collisions in the resulting dataframe.

```
# first dataframe
x = [1,2,3]
y = ['a','b','c']
g = [1,2,3]
h = ['Z','Y','Y']
df = pd.DataFrame(zip(x,y,g,h),
                  columns=['x','y','g','h'])

# second dataframe only has 1/2
v2 = [10,11]
g2 = [1,2]
h2 = ['Z','Z']
df2 = pd.DataFrame(zip(v2,g2,h2),columns=['v2','g','h'])

# not matching on h, will rename
dfm = df.merge(df2,on='g',how='inner')
print(dfm)
```

```
   x  y  g h_x  v2 h_y
0  1  a  1   Z  10   Z
1  2  b  2   Y  11   Z
```

You can also merge on keys that are not named the same between the two dataframes.

```
# first dataframe
x = [1,2,3]
y = ['a','b','c']
g = [1,2,3]
df = pd.DataFrame(zip(x,y,g),
                  columns=['x','y','g'])

# second dataframe now has a new 4th group
v2 = [10,11,12]
g2 = [1,2,4]
df2 = pd.DataFrame(zip(v2,g2),columns=['v2','g2'])

# merging different var names
dfm = df.merge(df2,left_on='g',right_on='g2',
               how='outer')
print(dfm)
```

```
     x    y    g    v2   g2
0  1.0    a  1.0  10.0  1.0
1  2.0    b  2.0  11.0  2.0
2  3.0    c  3.0   NaN  NaN
3  NaN  NaN  NaN  12.0  4.0
```

I have shown an outer join here, so the `g=4` from the second dataframe is included in the results.

The final type of join I want to show is a cross join. This creates a cartesian product between each dataset, so if one dataset has 4 rows, and the other has 3 rows, the final result will be 12 rows:

```
# first dataframe 4 rows
x = [1,2,3,4]
y = ['a','b','c','d']
df = pd.DataFrame(zip(x,y),
                  columns=['x','y'])

# no same keys, 3 rows
v2 = [10,11,12]
g2 = ['Z','Y','X']
df2 = pd.DataFrame(zip(v2,g2),columns=['v','g'])

# merging different var names
dfm = df.merge(df2,how='cross')
print(dfm)
```

```
    x  y   v  g
0   1  a  10  Z
1   1  a  11  Y
2   1  a  12  X
3   2  b  10  Z
4   2  b  11  Y
5   2  b  12  X
6   3  c  10  Z
7   3  c  11  Y
8   3  c  12  X
9   4  d  10  Z
10  4  d  11  Y
11  4  d  12  X
```

You can use keys in a cross join (or a many to many join). For a common crime analysis example, imagine we want to create pairs of individuals for a social network analysis that were stopped together. You can cross join a table to itself to create the edge table for social network analysis.

```
# first dataframe 4 rows
IncID = [1,1,2,2,2]
PersID = ['a','b','a','c','d']
df = pd.DataFrame(zip(IncID,PersID),
                  columns=['Inc','Pers'])

# join to itself
dfc = df.merge(df.copy(),on='Inc')

# eliminate duplicates, order does not matter
dfc = dfc[dfc['Pers_x'] < dfc['Pers_y']]
print(dfc)
```

```
   Inc Pers_x Pers_y
1    1      a      b
5    2      a      c
6    2      a      d
9    2      c      d
```

The other type of operation is unioning two datasets together (stacking them on top of one another). To do this, you would use `pd.concat`:

```
# first dataframe year, 2023
y = [2022]*2
x = [1,2]
df22 = pd.DataFrame(zip(x,y),columns=['x','y'])
```

```
y = [2023]*2
x = [1,2]
z = ['a','b']
df23 = pd.DataFrame(zip(x,y,z),columns=['x','y','z'])

# This stacks/unions the dataframes
# on top of one another
dflist = [df22,df23]
dfc = pd.concat(dflist)
print(dfc)
```

```
   x     y    z
0  1  2022  NaN
1  2  2022  NaN
0  1  2023    a
1  2  2023    b
```

Note that `df23` has an additional column. This is included in the stacked result, but is missing for the first 2 rows. Note also the index is stacked from each dataframe. Often I want to reset the index, which you can do `pd.concat(dflist,ignore_index=True)`.

If you have many dataframes and you want to merge them together, *and* they have the same indices (so you want to merge columns, not stacked rows), you can use `pd.concat(dflist,axis=1)`.

One last example I want to show in this section is *reshaping* data from wide to long and long to wide. Sometimes we have on a single row multiple attributes – pretend it is crime incident data, and you have multiple columns corresponding to the 1st reporting officer, 2nd reporting officer, etc. This is *wide* format. You may wish to reshape the data to *long* format, as the analysis you want to conduct is easier in that reshaped data.

```
# wide format
id = [1,2,3]
off1 = ['a','b','c']
off2 = [None,'d','e']
e = [5,9,1]
df = pd.DataFrame(zip(id,off1,off2,e),
     columns=['ID','Off1','Off2','extra'])

# reshaping wide to long
df_long = pd.wide_to_long(df,
          stubnames="Off",
          i="ID", j="Officer")
print(df_long)
```

```
            extra   Off
ID Officer
1  1            5      a
2  1            9      b
3  1            1      c
1  2            5   None
2  2            9      d
3  2            1      e
```

You can pass in multiple variables to stubnames, e.g. if you have `Name1` and `Name2` as officer names, you could do `stubnames=['Off','Name'],j=['Officer','Name']` and both of the variables would be reshaped from wide to long.

Sometimes you want to go in the opposite direction, pretend we have a dataset of charges per incident in long format, and you want to reshape so that each row is a single crime incident. You can do that via `pivot`.

```
# long format
id = [1,1,1,2,2]
charge = ['rob','assault','weap','theft','contra']
ord = [1,2,3,1,2]
e = [9,1,2,5,3]
df = pd.DataFrame(zip(id,charge,ord,e),
     columns=['ID','charge','ord','extra'])

# reshaping wide to long
df_wide = df.pivot(index='ID',
                   columns='ord',
                   values=['charge','extra'])
print(df_wide)
```

```
    charge                  extra
ord      1        2     3     1  2    3
ID
1      rob  assault  weap     9  1    2
2    theft   contra   NaN     5  3  NaN
```

You can see in this example that the column names have multiple levels. If you want to collapse them so they are not as complicated, you can do something like:

```
# renaming column names
df_wide.columns = [f"{n}_{i}" for n,i in df_wide.columns]
print(df_wide)
```

```
   charge_1 charge_2 charge_3 extra_1 extra_2 extra_3
ID
1       rob  assault     weap       9       1       2
2     theft   contra      NaN       5       3     NaN
```

Also note that since ID1 had three charges, and ID2 only had two charges, when reshaped to long, ID2 has missing data for the third column values. This will reshape to whatever the max subcolumn is.

6.7 Looping and Functions

To loop over a dataframe, you may think that something like below would work.

```
x = [1,2,3]
y = ['z','y','x']
df = pd.DataFrame(zip(x,y),columns=['x','y'])

# naive loop probably does not do what you want
for i in df:
  print(i)
```

```
x
y
```

This just prints `x` then `y` though – the naive loop loops over *column* names. If that is what you want, I would suggest using `for col in df.columns` – that makes the code much more explicit. And there may be scenarios you want to do that – say you wanted to strip all text fields in your dataframe, you may do something like below:

```
x = [1,2,3]
y = [' z',' y ','x  ']
df = pd.DataFrame(zip(x,y),columns=['x','y'])
# can see the whitespace
print(df)

# Checking if string, if yes, strips extra whitespace
for col,dtype in zip(df.columns,df.dtypes):
  # checking if a string object
  if dtype == object:
    df[col] = df[col].str.strip()

print('\nWhitespace is now stripped')
print(df)
```

```
   x    y
0  1    z
1  2   y
2  3  x

Whitespace is now stripped
   x  y
0  1  z
1  2  y
2  3  x
```

Because we have a mix of numeric and string fields, if you just did `df[col].str.strip()` on every field, it would fail on the numeric fields. But because we do a check first, it only attempts the strip operation when we have checked that the field type is `object` first (pandas form of a string field).

> Note
>
> This example works, even if you have a mixed field type, e.g. has `np.nan` for missing in some of the fields. But to be defensive, you may wrap the code in a try-except block.
>
> ```
> # wrapping in try/except
> for col,dtype in zip(df.columns,df.dtypes):
>   # checking if a string object
>   if dtype == object:
>     try:
>       df[col] = df[col].str.strip()
>     except:
>       pass
> ```
>
> If this fails, it will just move onto the next column. This is useful when using functions that may cause errors if the field does not exactly conform to what you expect.

To iterate over the rows of a pandas dataframe, you use `df.itertuples()`.

```
x = [1,2,3]
y = ['z','y','x']
df = pd.DataFrame(zip(x,y),columns=['x','y'])

# looping over rows
for row in df.itertuples():
  print(row)
```

```
Pandas(Index=0, x=1, y='z')
Pandas(Index=1, x=2, y='y')
Pandas(Index=2, x=3, y='x')
```

You can see that each row prints out `Pandas(Index=0,x=1,y='z')` – these are equivalent to an object I have not introduced so far, *named tuples*. To access the specific named items in a named tuple, you can type `row.x` (for the x variable), or you can use `row[0]` for the first column.

For a simple example, imagine you only wanted to apply a function under certain circumstances – if a value is in a particular list, strip it from the dataframe (something you may do to protect privacy for rare values), otherwise leave it alone.

```
x = [1,2,3]
y = ['z','y','x']
df = pd.DataFrame(zip(x,y),columns=['x','y'])

# list to check
check_list = ['z','a']
res = [] # placing the results back into here

# looping over rows
for row in df.itertuples():
  if row.y in check_list:
    res.append('-')
  else:
    res.append(row.y)

# add back into the dataframe
df['y'] = res
print(df) # first row should have z replaced
```

```
   x  y
0  1  -
1  2  y
2  3  x
```

For this example, you could have equivalently done `for y in df['y']:` – loop over the series object directly.

The pattern here is fairly complicated – you need to place the results of the transformation into a new object (here a list), then add them back in. In this approach you cannot modify the item in place. To do that, you could loop over the indices directly, and use `df.loc` to modify the item inplace.

```
x = [1,2,3]
y = ['z','y','x']
df = pd.DataFrame(zip(x,y),columns=['x','y'])

# list to check
check_list = ['z','a']
```

```
# looping over the indices
for row in df.index:
  if df.loc[row,'y'] in check_list:
    df.loc[row,'y'] = '-'

print(df) # modified inplace
```

```
   x  y
0  1  -
1  2  y
2  3  x
```

I have shown these examples of looping over columns or rows, but often you want to strive to not loop at all. For this simple replace scenarios, you could have done the dictionary replace to accomplish the same thing:

```
x = [1,2,3]
y = ['z','y','x']
df = pd.DataFrame(zip(x,y),columns=['x','y'])

# list to check
check_list = ['z','a']

# creating dictionary to replace
rep_di = {c:'-' for c in check_list}

df['y'] = df['y'].replace(rep_di)
print(df) # easier than looping
```

```
   x  y
0  1  -
1  2  y
2  3  x
```

For datasets with many rows, this will be *much faster* than the loop approach. (Another function to check out is `df.where`, if you have a binary condition, you can replace items in just the locations that meet or fail to meet the condition.) For more complicated functions, you can use a method to apply that function to every row in the dataset. Here is an example of doing that:

```
x = [1,2,3]
y = ['z','y','x']
df = pd.DataFrame(zip(x,y),columns=['x','y'])
```

```
# mask function
def mask(x,check_list=['z','a'],rep='-'):
  if x in check_list:
    return rep
  else:
    return x

# can apply this to every row
df['y'] = df['y'].apply(mask)
print(df)
```

```
   x  y
0  1  -
1  2  y
2  3  x
```

You can pass additional keyword args to the function, so if you wanted to replace the string with a "*" instead of a dash, you could do `df['y'].apply(mask,rep="*")`.

> Note
>
> Some examples you see of this will use *lambda* functions. They are just ways to define a function inline. For example, if you had a complicated numeric function, you could do something like:
>
> ```
> # complicated one liner lambda apply
> df['sqn_div3'] = df['n'].apply(lambda x: (x**2)/3)
> ```
>
> So instead of having a function defined in the global scope, it is just defined directly in the apply function call. I don't tend to personally use these, as scenarios where I need to define a new function and they are simple enough to fit on a single line are rare in my experience.

A final example with apply I want to show is that apply functions can take multiple arguments across rows or columns – so you use `apply` on a dataframe, not on a series.

Here I use a simple example for a range function, I take the difference in the min and max – the range of values.

```
x = [1,2,3]
y = ['a','b','c']
z = [6,5,4]
df = pd.DataFrame(zip(x,y,z),columns=['x','y','z'])

# max - min
def ra_val(x):
  return max(x) - min(x)
```

```
# can apply this to every row
df['ra'] = df[['x','z']].apply(ra_val,axis=1)
print(df)
```

```
   x  y  z  ra
0  1  a  6   5
1  2  b  5   3
2  3  c  4   1
```

I show a few different things here. One, I select out only the variables I want to be evaluated in the function via `df[['x','z']]`. If I just did `df.apply(ra_val,axis=1)`, it would pass in a string field, and the function would return an error.

Two, I pass in the argument `axis=1`. This tells the apply function to apply it across the rows. If you do `axis=0`, it applies the function across columns, and would only return two results, one for x and one for z:

```
# applying to columns
col_ra = df[['x','z']].apply(ra_val,axis=0)
print(col_ra)
```

```
x    2
z    2
dtype: int64
```

So the result cannot be added back into the original dataframe (note the default keyword arg for apply to dataframes is `axis=0`).

The way that arguments are passed to this function are via a list. Here is an example taking in two string arguments and manipulating them:

```
x = ['Z','y','X']
y = ['a','b','c']
df = pd.DataFrame(zip(x,y),columns=['x','y'])

# combining strings
def comb_str(x):
    sc = x[0].lower() + "_" + x[1].upper()
    return sc

# order here matters!
df['cs'] = df[['y','x']].apply(comb_str,axis=1)
print(df)
```

```
   x  y   cs
0  Z  a  a_Z
1  y  b  b_Y
2  X  c  c_X
```

You can additionally return a list via a function. By default when you do that, you have a list stuffed into each row:

```
x = ['Z','y','X']
y = ['abc','bcd','cde']
df = pd.DataFrame(zip(x,y),columns=['x','y'])

# several string funcs
def str_clean(x):
    res = []
    for i in x:
        inew = i.title()
        inew = inew.strip()
        inew = inew + "_"
        res.append(inew)
    return res

# can apply this to every row
df['cs'] = df.apply(str_clean,axis=1)
print(df)
```

```
   x    y          cs
0  Z  abc  [Z_, Abc_]
1  y  bcd  [Y_, Bcd_]
2  X  cde  [X_, Cde_]
```

But it may be more convenient to return these as new columns. For example you can replace many different columns at once by specifying the kwargs `result_type='expand'`.

```
# placing in a list
col_app = ['x','y']

# can distribute function calls
# on multiple rows, looks for end parens
res_expand = df[col_app].apply(str_clean,
                result_type='expand',axis=1)

print(res_expand)
```

```
   0     1
0  Z_  Abc_
1  Y_  Bcd_
2  X_  Cde_
```

You can add the results back into the dataframe directly all at once in this approach as well:

```
# can add in multiple columns
# at once into
df[col_app] = df[col_app].apply(str_clean,
              result_type='expand',axis=1)

print(df)
```

```
    x     y          cs
0  Z_  Abc_  [Z_, Abc_]
1  Y_  Bcd_  [Y_, Bcd_]
2  X_  Cde_  [X_, Cde_]
```

These particular examples of applying functions are very simple, but the power of being able to apply functions in this manner is that they can be whatever complicated function you need them to be. Generally a good code practice is to have your functions in a `src` folder, import them, and have your main script be less cluttered. Using apply on various data cleaning tasks can help with that.

Or simply having a function that is `clean_data = data_clean(orig_df)`. So it places *all* of the data manipulation stages you want into a single function call.

6.8 Reading in Local Data

So far in the chapter, I have used cheeky examples of small datasets, defined entirely in code, to illustrate various pandas functions. In real life, you will be reading in data from some external source, such as a csv or excel file, or a database.

Here is an example of reading in a csv file (csv stands for *comma separated values*) from a location on your machine (note you need to change the path to wherever your data is located).

```
# hypothetical read data from local computer
df = pd.read_csv('G:\PathToFile\local.csv')
```

You can do similar for excel files via `pd.read_excel` (for those you need to have additional external libraries installed to read the excel files, such as the *openpyxl* python library). You can additionally read in example datasets from web urls directly. Below for example reads in a dataset of the number of sworn and civilian employees for police departments, reported to the FBI, for 2022 I have posted online.

```
# long url, splitting across
# multiple lines to make it easier to read!
url = ('https://raw.githubusercontent.com/apwheele/'
       'apwheele/main/Officers2022.csv')

# reading in a data file from internet
df = pd.read_csv(url)

# description of counts
print(df[['officer_ct','civilian_ct']].describe())
```

```
        officer_ct   civilian_ct
count  14047.000000  14047.000000
mean      50.086567     22.021713
std      369.882043    169.803031
min        1.000000      0.000000
25%        6.000000      1.000000
50%       14.000000      3.000000
75%       34.000000     12.000000
max    34012.000000  14572.000000
```

Note

A useful trick, in addition to knowing `.describe`, is to use it in combination with groupby. I often do this to generate statistical summaries per group:

```
# can use groupby and describe at the same time
res_sum = (df.groupby('agency_type_name')['officer_ct'].
           describe())
print(res_sum)
```

Check what these results look like for yourself! Sometimes it is also useful to transpose this result, e.g. `res_sum.T`, for very long tables.

The following chapter shows a more common example for crime analysts though – reading in data from database tables using SQL.

7 An introduction to SQL

The prior chapter discussed how to work with data tables and conduct regular types of analysis with those tables. At the end of the chapter, I showed an example of how to load data into a python pandas dataframe from a csv file that is saved on your local computer. In practice, crime analysts will be reading in data *from a database*. Typically police departments have a *records management system* (RMS), that contains different tables of standard information. This includes tables like 911 calls, reported incidents, arrests, etc.

SQL, *Structured Query Language*, is the main mechanism analysts use to interact with databases. Pretty much all databases have a standardized language in which you can read tables, manipulate data, and merge tables. This chapter will show how to write SQL code to pull in data into python. This is a very important skill to have for all analysts (not just crime analysts). So this chapter illustrates using SQL in python, but will be a very good introduction to SQL itself in general.

> Note
>
> Although SQL is very standardized, there do tend to be differences in queries and functions across different databases. That is, a query may work in a SQL Server database, but not in Microsoft Access. I will attempt to highlight when I use SQL that may vary across different databases.

7.1 Initializing a database for the Chapter Examples

This first section creates an example database to illustrate SQL. You will need to run the code below (in your current python session), to be able to follow along with the other examples in the chapter. The below code initializes a database connection to an in-memory SQLite database. I will be using these tables in examples throughout the chapter. The table `Officers` contains information on the total number of sworn and non-sworn personal for different police agencies in the United States. And the file `Crimes` contains statistics on UCR crime reporting (both collected by the FBI) in 2022. Agencies are uniquely identified in each table by their *ORI*, originating agency identifier.

> Note
>
> This code uses a method to download data from the internet directly into a pandas dataframe. So if you do not have internet you will need to download these files somewhere else, like on a usb drive, and then save them on your local computer. When following along, you would read in those local csv files like I showed at the end of the prior chapter, e.g. something like `pd.read_csv(r'C:\local_folder\Crime2022.csv')`.

Don't worry if you don't know what "in-memory" means, or what "SQLite" is. SQL code is very standard across different databases, so the things I am going to show here will work if you replace `con` in the below script with a connection to your own database that you are working with. (And I show later in the chapter how to make those types of connections.) Using SQLite is a simple, portable way to help you get started learning how to run SQL queries in python.

```
import pandas as pd
import sqlite3

# create an in-memory database
con = sqlite3.connect(":memory:")

# read in data, add into database
ourl = ('https://github.com/apwheele/apwheele/raw/main/'
        'Officers2022.csv')
officer = pd.read_csv(ourl)
officer.to_sql('Officer',con=con,index=False)

# read in crimes, add into database
curl = ('https://github.com/apwheele/apwheele/raw/main/'
        'Crime2022.csv')
crimes = pd.read_csv(curl)
crimes.to_sql('Crime',con=con,index=False)

# showing size of each table
print(officer.shape, crimes.shape)
```

```
(14047, 17) (14375, 12)
```

For background on these tables, they are based on information disseminated by the FBI. The first is a table of officer sworn counts, and the second is a table of URC crime counts. So each row is reported metrics for a single police department in 2022. See https://cde.ucr.cjis.gov/ for a list of the tables.

Now that these tables are defined, you can create a SQL string, and then pass this to a pandas method `.read_sql`, to read the table from the database back into an in-memory object in python.

```
# reading in the crime table, displaying the results
query = 'SELECT * FROM Crime;'
crime_table = pd.read_sql(query,con)

# print first 3 rows transposed
print(crime_table.head(3).T)
```

```
                              0          1          2
ori                   OK0081500  NH0061800  MN0480400
aggravated_assault            4          0          8
arson                         0          0          0
burglary                      4          0          3
homicide                      0          0          0
human_trafficing              0          0          0
larceny                      24          3         17
motor_vehicle_theft          10          1          2
property_crime               38          4         22
rape                          0          2          0
robbery                       1          0          0
violent_crime                 5          2          8
```

This uses the `con` object, and a string that represents a query, to read in the table into a pandas dataframe in your python session.

> Note
>
> A common question crime analysts have is "how do I connect with my RMS database to run queries?". You need to ask individuals, either who manage the RMS or IT services for the department, where the database is located. A later section in the chapter gives more examples of connecting to different databases using additional python libraries.

7.2 SQL SELECT basics

The prior example had the query `SELECT * FROM Crimes;`. In SQL, unlike python, white space does not matter, and case most of the time does not matter. So the command could be equivalently written as `select * from crimes;`. To make it easier to read SQL though, I typically have the conventions of using ALL CAPS for SQL commands, lowercase for database fields, and Title case for database tables (case sometimes matters with field names and table names, but many databases force you to keep these in lower case). I also attempt to use whitespace to delimit different parts of the SQL, such as field name selection and other transformations, but unlike in python code, whitespace and line breaks in SQL are arbitrary.

The semi-colon is the terminating character for a query. It is often not needed when running SQL code from python, as many queries in python are limited to only executing one statement at a time. So in python you cannot run `SELECT * FROM Table1; SELECT * FROM Table2;` and get back two tables, you would need to run the queries separately one at a time.

The way to read the `SELECT * FROM Crimes;` query in plain English is "I want to select all fields from the crimes table". The asterisk symbol, `*`, signifies you want to query all of the fields at once. If you only want to select certain fields, you can specifically select the fields, and separate them by a comma.

```
# reading in the crime table, displaying the results
query = '''
-- This is a comment line in SQL
SELECT
  ori,
  violent_crime
FROM Crime
'''
# note no semi-colon, but this still works

crime_table = pd.read_sql(query,con)
# Now only has two fields
print(crime_table.head(3))
```

```
         ori  violent_crime
0  OK0081500              5
1  NH0061800              2
2  MN0480400              8
```

There are a few things to note here, in addition to only specifying two specific variables, here the ORI and violent crime counts, instead of the full set of fields from the table. First, I did this query on multiple lines, using the triple quoted, `'''`, string format I showed in Chapter 3. I could have done it on a single line, but in my opinion it is easier to read the query when spread out on multiple lines.

I also inserted a comment line, starting with two dashes, `-- comment line in SQL`. I also added in spaces before the field names. This is not necessary in SQL, but again I do it because it makes it easier to read.

The last thing I want to note in this example, is that there is no comma *after* `violent_crime`. That is a common mistake many people (including myself!) make when constructing SQL queries. You need commas between field names, but there should be no comma before the `FROM` command.

> Note
>
> Although it is often convenient to query all fields in a table using the `*` shorthand, it is typically good practice to only query the data you need. This means only selecting the fields and the rows you need to conduct the analysis at hand. This becomes especially important when working with large tables that may not fit into your computers RAM, random access memory. This tends to only be a problem for tables with millions of rows and dozens of variables though.

These examples limited the variables that were returned, you can also limit the rows. You can write a query to only return a certain number of rows by using `LIMIT`:

```
query = '''
/* This only returns 3 rows of data
```

```
    and this is a multi-line comment */
SELECT
  ori,
  violent_crime
FROM Crime
LIMIT 3
'''

crime_table = pd.read_sql(query,con)
# Should be limited to 3 rows
print(crime_table)
```

```
         ori  violent_crime
0  OK0081500              5
1  NH0061800              2
2  MN0480400              8
```

This is not very useful as is, because it just returns whatever the first three rows in the data table happen to be (which can be quite random). You can however order the data, to say return the agencies with the most recorded violent crime via an `ORDER BY` statement.

```
query = '''
SELECT
  ori,
  violent_crime
FROM Crime
ORDER BY violent_crime DESC
LIMIT 3;
'''

crime_table = pd.read_sql(query,con)
# This will return the Top 3 highest violent crime counts
print(crime_table)
```

```
         ori  violent_crime
0  NY0303000          61293
1  CA0194200          31772
2  TXHPD0000          25987
```

> Note
>
> Some different databases, instead of `LIMIT`, use a `SELECT TOP 3 * FROM Table;` like syntax. Access or Teradata are like this for example, and do not use the `LIMIT` syntax.

This works well when you only want a specified number of results returned, for example the top 10 high crime street segments in your jurisdiction to help with a hotspots policing program. But a likely more common example is to limit the rows based on certain criteria, such as greater than a value, or within a particular group. Here for example, I limit the result for the *officers* table to jurisdictions who are serving over 1 million population. You can do this via the `WHERE` command.

```
query = '''
SELECT
  ncic_agency_name,
  population
FROM Officer
WHERE population > 1000000;
'''

off_table = pd.read_sql(query,con)
# Only cities more than 1 million pop
print(off_table)
```

```
                          ncic_agency_name  population
0                               CHICAGO PD     2652124
1                 DALLAS POLICE DEPARTMENT     1286121
2     LAS VEGAS METROPOLITAN POLICE DEPARTMENT  1667961
3                PHOENIX POLICE DEPARTMENT     1637902
4             MONTGOMERY COUNTY PD ROCKVILLE   1030232
5   PHILADELPHIA POLICE DEPARTMENT POLICE ADMIN 1555812
6                        FAIRFAX COUNTY PD     1094682
7                             SAN DIEGO PD     1377838
8                HOUSTON POLICE DEPARTMENT     2276533
9                     HARRIS CO SO HOUSTON     2029908
10                          PD SAN ANTONIO     1465608
11               NEW YORK CITY PD NEW YORK     8236567
12               SUFFOLK COUNTY PD YAPHANK     1355324
13                 HILLSBOROUGH CO SO TAMPA    1052613
14                NASSAU COUNTY PD MINEOLA     1087353
15                          LOS ANGELES PD     3809182
```

Here you can do the typical `<` for less than, or `<=` for less than or equal to. You can also do equality constraints (makes more sense for categorical variables), or not equal is `!=` or `<>`.

```
query = '''
SELECT
  ncic_agency_name,
  population
FROM Officer
WHERE population > 1000000
  AND state_abbr = 'TX';
'''

off_table = pd.read_sql(query,con)
# Only cities more than 1 million pop in Texas
print(off_table)
```

```
          ncic_agency_name  population
0   DALLAS POLICE DEPARTMENT     1286121
1  HOUSTON POLICE DEPARTMENT     2276533
2       HARRIS CO SO HOUSTON     2029908
3             PD SAN ANTONIO     1465608
```

Here I combined multiple `WHERE` clauses using `AND`. You can combine `WHERE` clauses similarly using `OR` as well.

```
query = '''
SELECT
  ncic_agency_name,
  population
FROM Officer
WHERE population > 1000000
  AND (state_abbr = 'TX' OR state_abbr = 'CA');
'''

off_table = pd.read_sql(query,con)
# Only cities more than 1 million pop in Texas and California
print(off_table)
```

```
          ncic_agency_name  population
0   DALLAS POLICE DEPARTMENT     1286121
1               SAN DIEGO PD     1377838
2  HOUSTON POLICE DEPARTMENT     2276533
3       HARRIS CO SO HOUSTON     2029908
4             PD SAN ANTONIO     1465608
5            LOS ANGELES PD     3809182
```

When checking for multiple categories though in a single variable, the SQL command `IN (...list...)` can be very convenient.

```
query = '''
SELECT
  ncic_agency_name,
  population
FROM Officer
WHERE population > 1000000
  AND state_abbr IN ('TX','CA','NY');
'''

off_table = pd.read_sql(query,con)
# Only cities more than 1 million pop in
# Texas/California/New York
print(off_table)
```

```
             ncic_agency_name  population
0    DALLAS POLICE DEPARTMENT     1286121
1                SAN DIEGO PD     1377838
2   HOUSTON POLICE DEPARTMENT     2276533
3        HARRIS CO SO HOUSTON     2029908
4             PD SAN ANTONIO      1465608
5   NEW YORK CITY PD NEW YORK     8236567
6   SUFFOLK COUNTY PD YAPHANK     1355324
7    NASSAU COUNTY PD MINEOLA     1087353
8              LOS ANGELES PD     3809182
```

Note that you can negate this as well via `NOT IN (...)`, so if you did above and you wanted all states that *are not* a list, you could write:

```
AND state_abbr NOT IN ('TX','CA','NY')
```

A final good SQL trick to note is using the `LIKE` operation to return subsets of data that are not exact matches on strings. For example here, you likely don't know a PDs ori offhand, but you may search for a particular city by name. Here I do a search for `RALEIGH` anywhere in the agency name string. You do this via having the wildcard characters `%` placed before or after the string you are searching for.

```
query = '''
SELECT
  ori,
  ncic_agency_name
FROM Officer
WHERE ncic_agency_name LIKE '%RALEIGH%'
```

```
'''
off_table = pd.read_sql(query,con)
# agencies that have RALEIGH somewhere in name field
print(off_table)
```

```
          ori                                 ncic_agency_name
0   NCWLF0000              WILDLIFE PROTECTION DIV RALEIGH
1   NCNHP0000                        NC HP HQTRS RALEIGH
2   WV0410000                     RALEIGH CO SO BECKLEY
3   NC0922800        ST PARK RANGER UMSTEAD ST PARK RALEIGH
4   NC0929000     NC DIV OF ALCOHOL LAW ENFORCEMENT RALEIGH
5   NC0921200                   STATE CAPITOL PD RALEIGH
6   NC0922600   ST PARK RANGER FALLS LAKE ST REC AREA RALEIGH
7   NC0920100                                RALEIGH PD
8   NC0921500            RALEIGH-DURHAM AIRPORT PD RALEIGH
9   NC0922100  ST FAIRGROUNDS PUB SAFETY DEPT OF AGRI RALEIGH
10  NC0921600                NC STATE UNIV POLICE RALEIGH
11  NC0920000                        WAKE CO SO RALEIGH
12  MS0650000                       SMITH CO SO RALEIGH
```

The percent characters here allow for the string to have any number of characters before or after the word `RALEIGH`, `RALEIGH` just needs to be contained somewhere in the agency name field for it to be selected.

If you only wanted RALEIGH to be at the start of the string, you could have used `LIKE 'RALEIGH%'`, or if just at the end of the string `LIKE '%RALEIGH'`. The percent symbol indicates to pad with any number of characters – if you used `?` it looks for a single character in the string. So `LIKE '??? RALEIGH%` would only return strings with 3 characters, a space, and then RALEIGH in the string field.

> Note
>
> SQL has additional string field manipulations, like trimming white space and converting to upper/lower case. So if you have a field with mixed strings here, it may make sense to do `WHERE UPPER(agency_name) LIKE '%RALEIGH%'` to not have to worry about having the correct case in the like search.

7.3 Field Manipulations

In SQL, you can derive new values from current fields. For example you can *add* two fields together. Here I add the female sworn and civilian counts together to get the total females working in police departments.

```
query = '''
SELECT
  ori,
  female_officer_ct + female_civilian_ct AS calc_female,
  female_total_ct AS orig_female
FROM Officer
LIMIT 4
'''

off_table = pd.read_sql(query,con)
# Summed female field
print(off_table)
```

```
         ori  calc_female  orig_female
0  AL0010000          190          190
1  AL0010100           41           41
2  AL0010200          225          225
3  AL0010300            9            9
```

You can see that the calculated field is the same as the total female field already in the data table. The general format in SQL is `...field calculation... AS final_name`, so you use the `AS` keyword to determine what the field name will be for the column in the resulting table (if you omit `AS`, it will still return a result, but it can often be a confusing default name given). And here I show that you can rename original field names as well, `female_total_ct AS orig_female`, even if there is no calculation being performed.

For other numeric examples you can subtract, divide, multiply, as well as do multiple of these operations at once. You can do these operations either compared to other fields, or as specific constants you input into the SQL.

```
query = '''
SELECT
  female_total_ct AS fem_tot,
  female_officer_ct AS fem_off,
  female_total_ct - female_officer_ct AS calc_femciv,
  -- Can do a constant instead of different field
  female_total_ct - 100 AS min100_femtot,
  female_officer_ct/female_total_ct AS prop_fem
FROM Officer
LIMIT 4
'''

off_table = pd.read_sql(query,con)
# Different field calculations
```

```
print(off_table)
```

```
   fem_tot  fem_off  calc_femciv  min100_femtot  prop_fem
0      190       79          111             90         0
1       41       12           29            -59         0
2      225       99          126            125         0
3        9        6            3            -91         0
```

I have put a column in here that shows a problem, the calculated `prop_fem` column at the very end. For the first row, this should be `79/190`, which approximately equals `0.416`. But it shows all the values as 0 in this example, what gives? SQL by default when doing division of integers also by default creates the calculated column as an integer, so in these examples the result is always truncated down to 0.

To be able to get a proper proportion, you need to use the function `CAST` to change at least one of the input integers to a float value. If you cast one of the values (either the numerator or the denominator) to a float, the resulting calculation will be as you expect. Here I also multiply by 100, to get the resulting value on a percentage scale.

```
query = '''
SELECT
  female_total_ct,
  female_officer_ct,
  /* Casting to a float before the division
     can also do additional computations */
  100*CAST(female_officer_ct AS float)/female_total_ct
    AS perc_femsworn
FROM Officer
LIMIT 4
'''

off_table = pd.read_sql(query,con)
# Percent sworn females should now be correct
print(off_table)
```

```
   female_total_ct  female_officer_ct  perc_femsworn
0              190                 79      41.578947
1               41                 12      29.268293
2              225                 99      44.000000
3                9                  6      66.666667
```

You can also do operations on string fields. Here is an example of selecting portions of a string using the `SUBSTRING` function.

```
query = '''
SELECT
  ori,
  -- select first 2 characters
  SUBSTRING(ori,1,2),
  -- select last four characters
  SUBSTRING(ori,-4) AS end_4ori,
  -- get middle characters
  SUBSTRING(ori,3,3) AS mid_ori
FROM Officer
LIMIT 4
'''

off_table = pd.read_sql(query,con)
print(off_table)
```

```
         ori SUBSTRING(ori,1,2) end_4ori mid_ori
0  AL0010000                 AL     0000     001
1  AL0010100                 AL     0100     001
2  AL0010200                 AL     0200     001
3  AL0010300                 AL     0300     001
```

While many SQL operations are the same, some of the string functions do change across different databases. For example, Microsoft Access has a `LEFT`, `MID`, and `RIGHT` function (same as Excel), but these are not always available in other databases.

> Note
>
> Another way to get the end of the string is to calculate its length, and then get the last four characters. For example `SUBSTRING(ori,LENGTH(ori)-4,4)`. You can pass in other variables as arguments in the SQL string functions.

One can also concatenate string fields, here is an example of combining the agency name and the state:

```
query = '''
SELECT
  ori,
  -- combining string fields
  (agency_name_edit || ", " || state_abbr) AS comb_name
FROM Officer
LIMIT 4
'''

off_table = pd.read_sql(query,con)
```

```
print(off_table)
```

```
         ori                         comb_name
0  AL0010000  JEFFERSON CO SO BIRMINGHAM, AL
1  AL0010100                 BESSEMER PD, AL
2  AL0010200               BIRMINGHAM PD, AL
3  AL0010300           MOUNTAIN BROOK PD, AL
```

Some databases have a `CONCAT` function to do this as well, e.g. `CONCAT(field1,"|",field2)`, SQLite does not have a `CONCAT` function though. Note here how I inserted `", "` in-between the two fields. Without this comma and space, the result would be `JEFFERSON CO SO BIRMINGHAMAL` instead of `JEFFERSON CO SO BIRMINGHAM, AL`. If you need to be able to read the final result, you often need to include some type of delimiter and white space in-between the concatenated fields.

A final row level operation I want to illustrate is `CASE WHEN` – this is equivalent to if-else statements in python. You can check for certain conditions, and then assign a particular value when those conditions are met. Here for example I classify agencies according to their population size:

```
query = """
SELECT
  agency_name_edit,
  population,
  CASE
    WHEN population < 200000 THEN 'Small'
    WHEN population < 500000 THEN 'Middle'
    ELSE 'Large'
  END AS PopCat
FROM Officer
WHERE state_abbr = 'NC'
ORDER BY population DESC
LIMIT 10
"""

off_table = pd.read_sql(query,con)
print(off_table)
```

```
                            agency_name_edit  population  PopCat
0  CHARLOTTE-MECKLENBURG POLICE DEPARTMENT      955466   Large
1                               RALEIGH PD      470829  Middle
2                            GREENSBORO PD      298719  Middle
3                                DURHAM PD      286377  Middle
4                         WINSTON-SALEM PD      251295  Middle
5                        WAKE CO SO RALEIGH     210387  Middle
6                          FAYETTEVILLE PD      208980  Middle
7                                  CARY PD      178600   Small
8               JOHNSTON CO SO SMITHFIELD       172995   Small
9                     UNION CO SO MONROE        168556   Small
```

This example works since when a condition is met, it assigns a particular value. So I do not need to write `WHEN population >= 200000 AND population < 500000 THEN 'Middle'`, since I have written the conditions from smaller to larger. This is again very similar to if statements in pure python. One could have conditions that are not all related to the same column, e.g. you could have `CASE WHEN state_abbr NOT IN ('AK','HI') THEN 'Lower40'`, and for the second statements then have

```
WHEN population < 5000 THEN 'Tiny' ELSE 'Leftover' END
```

The if statements need not have any obvious relation to one another. And one can also nest multiple CASE WHEN statements as well.

There are additional string (and numeric) functions that I do not show in this chapter. For example, many databases will have the ability to change the case using `UPPER` or `LOWER` field functions. Some databases will have the ability to do regular expressions on fields. You can even make your own row level functions in databases as well. You will need to do your own research on your particular database to see what is available and what is possible.

7.4 Merging Tables

One of the main reasons to use a database is to define *relationships* between different tables. Say you want to join a table of reported crimes to arrests, to calculate the clearance rate of arrests over time. Or say you want to join a table of addresses with geo-coordinates to a table of crimes with addresses to be able to map out hot spots. These are again equivalent to operations I showed in the prior chapter using pandas merge, but can be accomplished in SQL as well.

```
query = """
SELECT
  c.ori,
  c.violent_crime,
  o.population,
  100000*CAST(c.violent_crime AS float)/o.population
```

```
    AS violent_rate
FROM Crime AS c
INNER JOIN Officer AS o
  ON c.ori = o.ori
LIMIT 10
"""

join_table = pd.read_sql(query,con)
print(join_table)
```

```
         ori  violent_crime  population  violent_rate
0  OK0081500              5        3144    159.033079
1  MN0480400              8         806    992.555831
2  IL0168800             65       20896    311.064319
3  MN0010400              0         620      0.000000
4  MI4620200            118       20407    578.232959
5  IA0100100              9        6236    144.323284
6  SD0090000              7        4938    141.757797
7  FL0310100             11        4978    220.972278
8  NC0420100             14        1838    761.697497
9  KY0160100              8        2459    325.335502
```

Here we match rows from the Crime table to the Officer table based on the matching key `ori`. We also for convenience give field aliases to the crime and officer tables – I rename crime to `c` and officer to `o`. You can then select fields from individual tables by referring to them via `o.field` and `c.field`. This is necessary when dealing with multiple tables to disambiguate the field names in the different tables.

The join type I use here is an *inner* join. This means that there needs to exist cases in both the crime and the officer table to have a row in the resulting table. If you do a left join, this means the table will contain all of the rows of the crime table, even if there are no associated rows in the officer table. (These are the same types of joins as I showed in the prior chapter using `pd.merge`.) Here you can see that some of the officer stats are missing when using a left join.

```
query = """
SELECT
  c.ori,
  c.violent_crime,
  o.population,
  100000*CAST(c.violent_crime AS float)/o.population
    AS violent_rate
FROM Crime AS c
LEFT JOIN Officer AS o
  ON c.ori = o.ori
LIMIT 10
```

```
"""

join_table = pd.read_sql(query,con)
print(join_table)
```

```
         ori  violent_crime  population  violent_rate
0  OK0081500              5      3144.0    159.033079
1  NH0061800              2         NaN           NaN
2  MN0480400              8       806.0    992.555831
3  IL0168800             65     20896.0    311.064319
4  MN0010400              0       620.0      0.000000
5  MI4620200            118     20407.0    578.232959
6  IA0100100              9      6236.0    144.323284
7  SD0090000              7      4938.0    141.757797
8  MO0551000              1         NaN           NaN
9  FL0310100             11      4978.0    220.972278
```

There are also `RIGHT JOIN` versions in SQL, which would include all rows in the officer table, even if there were no corresponding ori's in the crime table.

> Note
>
> An *outer join* means it contains rows from both datasets, not shown as SQLite does not have outer joins (but I will show an example of replicating an outer join in the multiple queries at once section later on in the chapter).

These two tables have what is called a *one-to-one* relationship. For each table, each row has a unique ori number. SQL can do joins though for other relationships, such as one to many, many to one, or many to many. For example, say you wanted to compare officer per capita rates in large cities within the same state to each other, you can join a table to itself.

```
query = """
SELECT
  l.pe_ct_per_1000 AS left_rate,
  l.ncic_agency_name AS left_name,
  r.pe_ct_per_1000 AS right_rate,
  r.ncic_agency_name AS right_name
FROM Officer AS l
LEFT JOIN Officer AS r
  ON l.state_abbr = r.state_abbr
WHERE l.population > 500000
  AND r.population > 500000
  AND l.population < r.population
LIMIT 10
"""
```

```
join_table = pd.read_sql(query,con)
print(join_table)
```

```
   left_rate                left_name  right_rate  \
0       2.90  DALLAS POLICE DEPARTMENT        2.28
1       2.90  DALLAS POLICE DEPARTMENT        2.76
2       2.90  DALLAS POLICE DEPARTMENT        2.13
3       3.51        SAN FRANCISCO PD        14.73
4       3.51        SAN FRANCISCO PD         3.12
5       3.51        SAN FRANCISCO PD         1.62
6       3.51        SAN FRANCISCO PD         1.64
7       1.85           SACRAMENTO PD         2.10
8       1.85           SACRAMENTO PD        14.73
9       1.85           SACRAMENTO PD         3.12

                 right_name
0       HARRIS CO SO HOUSTON
1   HOUSTON POLICE DEPARTMENT
2            PD SAN ANTONIO
3  LOS ANGELES CO SO LOS ANGELES
4            LOS ANGELES PD
5              SAN DIEGO PD
6               SAN JOSE PD
7                 FRESNO PD
8  LOS ANGELES CO SO LOS ANGELES
9            LOS ANGELES PD
```

The final where condition of `l.population < r.population` prevents having an agency compared to itself, as well as prevents duplicate rows just with the agencies switched. This is a convenient trick to be able to build social network edge datasets, where individuals are connected such as via associated the same incident report.

These examples all show joining a table based on rows, when you want to stack datasets on top of one another, you would do that via a `UNION`. This example stacks the ori's from each dataset together. The flag field, which is just a scalar constant for each table, distinguishes that the first row comes from the Crime table, and the second row comes from the Officer table.

```
query = """
SELECT
  ori,
  1 AS flag
FROM Crime
UNION
SELECT
```

```
  ori,
  2 AS flag
FROM Officer
LIMIT 2
"""

union_table = pd.read_sql(query,con)
print(union_table)
```

```
        ori  flag
0  AK0010100     1
1  AK0010100     2
```

For the scalar constants, you could also do `'CrimeTable' AS flag` – that is you could use strings instead of numeric values.

7.5 Aggregate Functions

The pandas operation `groupby` is a name that likely comes from the `GROUP BY` operation in SQL. You can estimate aggregate quantities within groups. The typical operations that are available across all SQL dialects are `COUNT`, the number of rows in a group, `AVG`, the mean of the group, `SUM`, `MAX`, and `MIN`. And I bet you can guess what each of those last three aggregate functions does just by reading them!

```
query = """
SELECT
  SUBSTRING(ori,1,2) AS State,
  AVG(robbery) AS MeanRob,
  COUNT(robbery) AS TotN,
  SUM(robbery) AS SumRob,
  MAX(robbery) AS MaxRob,
  MIN(robbery) AS MinRob
FROM Crime
WHERE State IN ('NC','SC')
GROUP BY State
"""

# Aggregate robbery stats
# within North/South Carolina
agg_table = pd.read_sql(query,con)
print(agg_table)
```

```
  State   MeanRob  TotN  SumRob  MaxRob  MinRob
0    NC 14.627551   392    5734    1381       0
1    SC  7.925094   267    2116     228       0
```

The groupby variable should always be one of the variables selected before the `FROM` statement. If you did not include that variable, you would not be able to tell which row was which state.

> Note
>
> In this example, I refer to the `State` variable that I created before the `FROM` statement. This is not always valid, especially with table joins and more complicated SQL. Some SQL dialects let you refer to calculated fields in the order in which they were created, e.g. in Postgres you can do `WHERE $1 IN ('NC','SC')`. Other times, you need to rewrite the operation multiple times, e.g. `GROUP BY SUBSTRING(ori,1,2)`.

The order with which you need to write a SQL statement (for the most part) is:

1) select the fields you want
2) select the tables
3) specify any joins
4) where statements to limit rows
5) group by
6) order by and having
7) limit

I use "for the most part" as some of these can be swapped (3 and 4 are somewhat interchangeable), and later I will show subqueries, which only need to obey this order within the subquery part. But this is overall a useful way to think about how the order of operations should go in your SQL statement.

So the group by statement always needs to come *after* all of the joins and where statements. The group by statement will never be before those statements. Sometimes you want to filter the resulting dataset based on the aggregated field, but the where statement is based on the original table. To do the aggregate value filter, you use the `HAVING` statement (after the `GROUP BY` statement) to limit those rows.

```
query = """
SELECT
  SUBSTRING(ori,1,2) AS State,
  AVG(robbery) AS MeanRob,
  COUNT(robbery) AS TotN,
  SUM(robbery) AS SumRob,
  MAX(robbery) AS MaxRob,
  MIN(robbery) AS MinRob
FROM Crime
GROUP BY State
HAVING TotN > 500
"""
```

```
# Aggregate robbery stats
# for states with more than 500 ORIs
agg_table = pd.read_sql(query,con)
print(agg_table)
```

```
  State    MeanRob  TotN  SumRob  MaxRob  MinRob
0    CA  62.420708   763   47627    9124       0
1    MI   6.124370   595    3644    1398       0
2    NJ   7.858268   508    3992     469       0
3    NY  40.891386   534   21836   17433       0
4    OH   9.983633   611    6100    1549       0
5    PA   8.172699   967    7903    5763       0
6    TX  20.469961  1032   21125    6955       0
```

You can have multiple variables in a group by statement, e.g. `GROUP BY Category1, Category2`. Here I construct a categorical variable for large/small agencies, and do a groupby on those multiple categories.

```
query = """
SELECT
  state_abbr,
  CASE
    WHEN population < 20000 THEN 'Small'
    ELSE 'Large'
  END AS AgSize,
  AVG(pe_ct_per_1000) AS MeanOffRate,
  COUNT(*) AS TotN
FROM Officer
GROUP BY state_abbr, AgSize
LIMIT 10
"""

# Aggregate for two variables
agg_table = pd.read_sql(query,con)
print(agg_table)
```

```
  state_abbr AgSize  MeanOffRate  TotN
0         AK  Large     1.952500     4
1         AK  Small     6.787667    34
2         AL  Large     2.887377    61
3         AL  Small     4.014578   345
4         AR  Large     2.846842    38
5         AR  Small     3.384398   263
6         AZ  Large     2.360465    43
7         AZ  Small     4.103750    78
8         CA  Large     2.406575   254
9         CA  Small     5.656090   216
```

Here I also show using `COUNT(*)` instead of `COUNT(pe_ct_per_1000)`. When using the asterisk, it counts all rows, whereas when using `COUNT(variable)` it only counts valid (non-null) values. Another common way to write this is `COUNT(1)`, although this is equivalent to `COUNT(*)`.

There are not many examples of aggregating string variables, one is equivalent to concatenating strings for the group by variable. In SQLite, this is the `GROUP_CONCAT` function, although other databases use `STRING_AGG`:

```
query = """
SELECT
  state_abbr,
  MAX(LENGTH(ori)) AS ori_len,
  MIN(ori) AS min_ori,
  GROUP_CONCAT(SUBSTRING(ori,1,7),"|") AS concat_ori
FROM Officer
WHERE population > 1000000
GROUP BY state_abbr
"""

# Concatenating strings together in GROUP BY
agg_table = pd.read_sql(query,con)
print(agg_table)
```

```
  state_abbr  ori_len    min_ori                               concat_ori
0         AZ        9  AZ0072300                                  AZ00723
1         CA        9  CA0194200                          CA03711|CA01942
2         FL        9  FL0290000                                  FL02900
3         IL        9  ILCPD0000                                  ILCPD00
4         MD        9  MD0160400                                  MD01604
5         NV        9  NV0020100                                  NV00201
6         NY        9  NY0290000                  NY03030|NY05101|NY02900
7         PA        9  PAPEP0000                                  PAPEP00
8         TX        9  TX1010000  TXDPD00|TXHPD00|TX10100|TXSPD00
9         VA        9  VA0290100                                  VA02901
```

Here I also show that you can do a string transformation, such as `LENGTH`, and then do numeric operations. I also show `MIN(ori)` here, and that returns the minimum based on the string in alphabetical order, e.g. `090` would be before `11` in alphabetic order.

One last example of aggregation I want to show is calculating ranks within a grouping variable. This uses a somewhat different syntax, and is what is called a *window* function in SQL parlance.

```
query = """
SELECT
  ncic_agency_name,
  population,
  RANK() OVER (ORDER BY population DESC) AS pop_rank
FROM Officer
WHERE population > 1000000
"""

# Aggregate robbery stats
# for states with more than 500 ORIs
rank_table = pd.read_sql(query,con)
print(rank_table)
```

```
                              ncic_agency_name  population  pop_rank
0                       NEW YORK CITY PD NEW YORK     8236567         1
1                                 LOS ANGELES PD     3809182         2
2                                     CHICAGO PD     2652124         3
3                     HOUSTON POLICE DEPARTMENT      2276533         4
4                          HARRIS CO SO HOUSTON      2029908         5
5       LAS VEGAS METROPOLITAN POLICE DEPARTMENT     1667961         6
6                     PHOENIX POLICE DEPARTMENT      1637902         7
7    PHILADELPHIA POLICE DEPARTMENT POLICE ADMIN     1555812         8
8                               PD SAN ANTONIO       1465608         9
9                                  SAN DIEGO PD      1377838        10
10                     SUFFOLK COUNTY PD YAPHANK     1355324        11
11                      DALLAS POLICE DEPARTMENT     1286121        12
12                             FAIRFAX COUNTY PD     1094682        13
13                      NASSAU COUNTY PD MINEOLA     1087353        14
14                      HILLSBOROUGH CO SO TAMPA     1052613        15
15                 MONTGOMERY COUNTY PD ROCKVILLE    1030232        16
```

You specify the rank via the order by clause after the `OVER` statement. You can also do this within an additional grouping variable, using the `PARTITION BY` clause:

```
query = """
SELECT
  ncic_agency_name,
  state_abbr,
  population,
  RANK() OVER (PARTITION BY state_abbr
               ORDER BY population DESC) AS pop_rank
FROM Officer
WHERE population > 1000000
  AND state_abbr IN ('TX','NY','CA')
"""

# Aggregate robbery stats
# for states with more than 500 ORIs
rank_table = pd.read_sql(query,con)
print(rank_table)
```

```
            ncic_agency_name state_abbr  population  pop_rank
0                LOS ANGELES PD         CA     3809182         1
1                  SAN DIEGO PD         CA     1377838         2
2  NEW YORK CITY PD NEW YORK         NY     8236567         1
3  SUFFOLK COUNTY PD YAPHANK         NY     1355324         2
4   NASSAU COUNTY PD MINEOLA         NY     1087353         3
5  HOUSTON POLICE DEPARTMENT         TX     2276533         1
6       HARRIS CO SO HOUSTON         TX     2029908         2
7             PD SAN ANTONIO         TX     1465608         3
8   DALLAS POLICE DEPARTMENT         TX     1286121         4
```

Window functions are also used when calculating lags and leads, for a hypothetical example `LAG(variable) OVER (ORDER BY time_period)` would get the prior value of `variable` ordered by the time period. And `PARTITION BY` works in a similar fashion for `LAG` as well.

7.6 Multiple Queries at Once

Imagine you wanted the highest population city in each state. To do that we can use a *subquery*.

```
query = """
SELECT * FROM (SELECT
  ncic_agency_name,
  state_abbr,
  population,
  RANK() OVER (PARTITION BY state_abbr
               ORDER BY population DESC) AS pop_rank
FROM Officer
WHERE state_abbr IN ('TX','NY','CA'))
WHERE pop_rank = 1
"""

# sub-query for top rank within each state
sub_table = pd.read_sql(query,con)
print(sub_table)
```

```
            ncic_agency_name state_abbr  population  pop_rank
0                LOS ANGELES PD         CA     3809182         1
1  NEW YORK CITY PD NEW YORK         NY     8236567         1
2  HOUSTON POLICE DEPARTMENT         TX     2276533         1
```

What is going on here? You can use a format of `SELECT * FROM (...another sql query...)`. That is, you can treat the section within the parentheses as a separate table. Sometimes this is useful to do selections, here I only select high population cities, and then merge in the crime statistics.

```
query = """
SELECT
  o.ori,
  o.ncic_agency_name,
  o.population,
  c.violent_crime
FROM (SELECT * FROM Officer WHERE population > 1000000) AS o
-- join by default is equivalent to an INNER JOIN
JOIN Crime AS c
  ON o.ori = c.ori
"""

# subquery for high population cities
sub_table = pd.read_sql(query,con)
print(sub_table)
```

```
          ori                               ncic_agency_name  population  \
0   ILCPD0000                                     CHICAGO PD     2652124
1   TXDPD0000                      DALLAS POLICE DEPARTMENT     1286121
2   NV0020100      LAS VEGAS METROPOLITAN POLICE DEPARTMENT     1667961
3   AZ0072300                     PHOENIX POLICE DEPARTMENT     1637902
4   MD0160400                 MONTGOMERY COUNTY PD ROCKVILLE     1030232
5   PAPEP0000  PHILADELPHIA POLICE DEPARTMENT POLICE ADMIN     1555812
6   VA0290100                             FAIRFAX COUNTY PD     1094682
7   CA0371100                                  SAN DIEGO PD     1377838
8   TXHPD0000                     HOUSTON POLICE DEPARTMENT     2276533
9   TX1010000                        HARRIS CO SO HOUSTON      2029908
10  TXSPD0000                               PD SAN ANTONIO     1465608
11  NY0303000                      NEW YORK CITY PD NEW YORK     8236567
12  NY0510100                      SUFFOLK COUNTY PD YAPHANK     1355324
13  NY0290000                       NASSAU COUNTY PD MINEOLA     1087353
14  CA0194200                                 LOS ANGELES PD     3809182

    violent_crime
0           14321
1           10009
2            8605
3           13515
4            1494
5           16202
6            1212
7            5932
8           25987
9            9199
10          12935
11          61293
12           1201
13           1748
14          31772
```

The end result is the same as if I did `WHERE o.population > 1000000` after joining the two tables. But in some circumstances, the subquery could be more efficient, as it could filter the Officer table first before attempting to do the join.

> Note
>
> SQL databases are very good at figuring out how to do complex operations on large tables. It is one of the reasons I have the SQL chapter, in many situations it is much easier to do computations in the database, and only work in python on the resulting already merged/aggregated table. Typically one would only need to worry about the subquery in situations with *tens of millions* of rows of data.

You can technically have nested sub-queries, but that can end up looking quite complicated. A simpler solution in that scenario is to use *common table expressions*, CTEs, to write more complicated queries with multiple tables. The above example rewritten with a CTE is below.

```
query = """
-- CTE to make table 1
WITH t1 AS (
  SELECT * FROM Officer WHERE population > 1000000
)

SELECT
  o.ori,
  o.ncic_agency_name,
  o.population,
  c.violent_crime
-- can now reference t1
FROM t1 AS o
JOIN Crime AS c
  ON o.ori = c.ori
"""

# Using CTE to get high pop cities
sub_table = pd.read_sql(query,con)
print(sub_table)
```

```
          ori                           ncic_agency_name  population  \
0   ILCPD0000                                 CHICAGO PD     2652124
1   TXDPD0000                  DALLAS POLICE DEPARTMENT     1286121
2   NV0020100     LAS VEGAS METROPOLITAN POLICE DEPARTMENT     1667961
3   AZ0072300                 PHOENIX POLICE DEPARTMENT     1637902
4   MD0160400             MONTGOMERY COUNTY PD ROCKVILLE     1030232
5   PAPEP0000   PHILADELPHIA POLICE DEPARTMENT POLICE ADMIN     1555812
6   VA0290100                         FAIRFAX COUNTY PD     1094682
7   CA0371100                              SAN DIEGO PD     1377838
8   TXHPD0000                  HOUSTON POLICE DEPARTMENT     2276533
9   TX1010000                      HARRIS CO SO HOUSTON     2029908
10  TXSPD0000                           PD SAN ANTONIO     1465608
11  NY0303000                  NEW YORK CITY PD NEW YORK     8236567
12  NY0510100                  SUFFOLK COUNTY PD YAPHANK     1355324
13  NY0290000                  NASSAU COUNTY PD MINEOLA     1087353
14  CA0194200                            LOS ANGELES PD     3809182

    violent_crime
0           14321
1           10009
2            8605
3           13515
4            1494
5           16202
6            1212
7            5932
8           25987
9            9199
10          12935
11          61293
12           1201
13           1748
14          31772
```

So the format here is generally:

```sql
WITH
  Table1 AS (...sql to create table...),
  Table2 AS (...sql that can reference Table1...)

-- need a final select to say what table to return
SELECT * FROM Table2;
```

You alias a table name after `WITH`, and can separate multiple CTEs with commas. Then you have a final SQL statement to produce the end result. The CTEs can reference other CTEs, or be totally independent. So you can chain together very complicated queries.

> Note
>
> One common database, Microsoft Access, does not allow CTEs, but does allow a limited set of nested sub-queries. You can however replicate the function of CTEs in Access by simply having multiple SQL statements (equivalent to VIEWS in other databases), and reference the intermediate views.

One example of using subqueries I want to show is a common pattern of checking existence in a separate table. For a prior example, I selected out cities in the officer table that had over 1 million population, and merged that result with the crimes table. But say I didn't need to merge in any info from the officer table, there is a simpler way using subqueries and `WHERE ... IN (..subquery...)`.

```
query = """
SELECT
  ori,
  violent_crime
FROM Crime
WHERE ori IN (
  SELECT ori FROM Officer
  WHERE population > 1000000)
"""

# WHERE in subquery, only big population cities
sub_table = pd.read_sql(query,con)
print(sub_table)
```

```
          ori  violent_crime
0   AZ0072300          13515
1   NY0303000          61293
2   NV0020100           8605
3   TX1010000           9199
4   TXSPD0000          12935
5   CA0371100           5932
6   NY0290000           1748
7   PAPEP0000          16202
8   MD0160400           1494
9   TXDPD0000          10009
10  CA0194200          31772
11  TXHPD0000          25987
12  ILCPD0000          14321
13  NY0510100           1201
14  VA0290100           1212
```

Say if you wanted to identify all incidents that are associated with some chronic offenders, and so you grab the IDs for the chronic offenders. This is like a filter on the larger table, based on a selection of certain characteristics in the subquery.

A final example of using CTEs I want to show is I mentioned SQLite does not have an outer join. The way I would accomplish an outer join of these two tables is using a CTE to develop a list of *all* the unique ori's across the two tables, and then merge in information from each of those tables. This uses a new type of query, `SELECT DISTINCT`, as well as `UNION`.

```
query = """
WITH ori_tab AS (
  SELECT DISTINCT ori FROM (
  SELECT ori FROM Crime
  UNION
  SELECT ori FROM Officer)
)

SELECT
  c.*,
  o.ncic_agency_name,
  o.population
FROM ori_tab
LEFT JOIN Crime AS c
  ON c.ori = ori_tab.ori
LEFT JOIN Officer as o
  ON o.ori = ori_tab.ori
"""

# distinct ORIs across both datasets
sub_table = pd.read_sql(query,con)
var_list = list(sub_table)
# limiting columns in output
print(sub_table[var_list[0:2] + var_list[-2:]].head(3))
```

```
         ori  aggravated_assault ncic_agency_name  population
0  AK0010100              2332.0     ANCHORAGE PD    285821.0
1  AK0010200               192.0     FAIRBANKS PD     32785.0
2  AK0010300                91.0        JUNEAU PD     31775.0
```

The first CTE, `SELECT DISTINCT`, stacks the two tables together, but eliminates rows that have duplicated ori's. And that CTE also shows having a subquery. A select distinct query is very similar to a set operation in python, it returns only unique results.

Then finally I use left joins to match the crime and officer table back to the complete list of ori's – you can have multiple joins in a single SQL query. Finally I show that with table aliasing, you can get *all* of the variables in a single table using `alias.*`, instead of selecting singular table values.

> Note
>
> Although using the `c.*` is convenient, note that if you did in the above example `c.*, o.*` you would have a *duplicate* ori field in the output. It returns a valid query, but having that duplicated field is unnecessary and can cause code headaches when you want to select out only a single field later on in python code. It is better in general to explicitly write out the variables you want in your SQL queries, even if it takes a bit more work.

7.7 Working with Dates

So far, we have been working with tables that were populated in the database via `df.to_sql('name',con)`, and then querying those tables via `pd.read_sql(query,con)`. Using the database connection object, you can execute arbitrary SQL code (as long as you have appropriate privileges), to do things like *create* new tables. Here I am going to show an example of creating a table, and populating it with some data, to help illustrate some date and time functions.

> Note
>
> Most crime analysts, when working with the *production* database that contains crime information will not have write privileges. I discuss later in the chapter a common approach to make a copy of the production database that the analyst has more control over.

```
create_dates = """
/* Creating table */
CREATE TABLE dt (
  id int,
  begin text,
  end text,
  crime text
);"""

# executing some arbitrary SQL
con.execute(create_dates)

# This table will be empty
print(pd.read_sql('SELECT * FROM dt',con))
```

```
Empty DataFrame
Columns: [id, begin, end, crime]
Index: []
```

The format for creating a table is `variable_name vartype`. In SQLite, there are no native datetime formats, but in different databases the type can be a specific date field. But often dates and times in databases are string fields, so you need to be aware that you can convert strings to dates in those circumstances.

Here I will insert two rows into the table. This is useful for testing purposes as I show here, but inserting in many values it is easier to use the `dataframe.to_sql` approach I showed at the beginning of the chapter.

```
insert_dates = """
-- Now inserting values, ISO dates
INSERT INTO dt (id,begin,end,crime)
VALUES
  (1,'2024-01-01 12:30:00','2024-01-01 12:40:00','Robbery'),
  (2,'2024-01-03 08:00:00','2024-01-03 16:00:00','Burglary');
"""

con.execute(insert_dates)

# This table will now have two rows
print(pd.read_sql('SELECT * FROM dt',con))
```

```
   id                begin                  end     crime
0   1  2024-01-01 12:30:00  2024-01-01 12:40:00   Robbery
1   2  2024-01-03 08:00:00  2024-01-03 16:00:00  Burglary
```

If you do `INSERT INTO` again, it will *append* rows to the table.

```
insert_dates = """
-- Now inserting values, ISO dates
INSERT INTO dt (id,begin,end,crime)
VALUES
  (3,'2024-01-04 12:30:00','2024-01-04 12:40:00','Robbery'),
  (4,'2024-01-03 08:00:00','2024-01-05 10:00:00','Burglary');
"""

con.execute(insert_dates)

# This table will now have four rows
print(pd.read_sql('SELECT * FROM dt',con))
```

```
   id                begin                  end     crime
0   1  2024-01-01 12:30:00  2024-01-01 12:40:00   Robbery
1   2  2024-01-03 08:00:00  2024-01-03 16:00:00  Burglary
2   3  2024-01-04 12:30:00  2024-01-04 12:40:00   Robbery
3   4  2024-01-03 08:00:00  2024-01-05 10:00:00  Burglary
```

Now this section I want to show some of the potential date and time manipulations. The begin and end fields are represented as strings in a particular date-time format. We can convert these fields to actual dates, and then calculate different date and time metrics on those fields. The way to do that in SQLite specifically is via converting the strings to `julianday`. Here I show creating a single row of data by passing in single scalar values and converting them to different date representations.

```
date_sql = """
SELECT
  julianday('2024-01-01') AS t1,
  julianday('2024-01-02') AS t2,
  julianday('2024-01-02') - julianday('2024-01-01')
    AS difftime;
"""

# showing creating dates
df = pd.read_sql(date_sql,con)
print(df)
```

```
          t1          t2  difftime
0  2460310.5  2460311.5       1.0
```

If you want sub-fractional days, you can either convert the fractional differences to your time period of choice, or use the `unixepoch` function to calculate seconds.

```
date_sql = """
SELECT
julianday('2024-01-02 12:30:00') -
julianday('2024-01-02 12:29:00')
  AS diffdays,
unixepoch('2024-01-02 12:30:00') -
unixepoch('2024-01-02 12:29:00')
  AS diffsecs
"""

# different days or seconds
df = pd.read_sql(date_sql,con)
print(df)
```

```
   diffdays  diffsecs
0  0.000694        60
```

You can convert days to seconds via `days*(60*60*24)`, that is there are 60 seconds in a minute, 60 minutes in an hour, and 24 hours in a day.

> Note
>
> SQLite is not the most convenient database to work with dates and times. If you are working with a different SQL database, you should check to see the date and time functions they have available. There is often a function named `DATEDIFF('days',EndDate,BeginDate)`, where you can place in different intervals such as hours, days, or months to calculate the time in between two different date fields.

If using our actual data, you can calculate the difference in dates and time directly on actual fields in the data.

```
date_sql = """
SELECT
  begin,
  end,
  julianday(end) - julianday(begin) AS diffdays
FROM dt;
"""

# difference between two fields
df = pd.read_sql(date_sql,con)
print(df)
```

```
                  begin                  end  diffdays
0  2024-01-01 12:30:00  2024-01-01 12:40:00  0.006944
1  2024-01-03 08:00:00  2024-01-03 16:00:00  0.333333
2  2024-01-04 12:30:00  2024-01-04 12:40:00  0.006944
3  2024-01-03 08:00:00  2024-01-05 10:00:00  2.083333
```

For crime events with large uncertainties you will want to conduct what is called *aoristic* analysis, see https://www.jratcliffe.net/aoristic-analysis for references. But for crime events with shorter uncertain intervals, calculating the midpoint is not so bad. To do that, we can calculate the difference between the begin and end, and then calculate a new field. Here I convert the result back into a string using the `datetime` function.

```
date_sql = """
SELECT
begin,
end,
datetime(
 unixepoch(begin) +
 (unixepoch(end) - unixepoch(begin))/2,
 'unixepoch') AS midtime
FROM dt;
"""
```

```
# calculating midpoint between two dates
df = pd.read_sql(date_sql,con)
print(df)
```

```
                 begin                  end              midtime
0  2024-01-01 12:30:00  2024-01-01 12:40:00  2024-01-01 12:35:00
1  2024-01-03 08:00:00  2024-01-03 16:00:00  2024-01-03 12:00:00
2  2024-01-04 12:30:00  2024-01-04 12:40:00  2024-01-04 12:35:00
3  2024-01-03 08:00:00  2024-01-05 10:00:00  2024-01-04 09:00:00
```

This is a bit complicated, so lets break it down into separate components. I calculate the difference in seconds between the end and begin times and divide that value by 2. I then add that value to the begin date. I then turn this back into an easier date field to read via the `datetime` function. This is the inverse function for `julianday` and `unixepoch`, it turns a numeric value back into a human readable string that represents the date and time fields.

A good trick to know in SQL is that you can use the string `'now'` to get *the current datetime* value. So if you wanted to calculate the days since an event has occurred (note this will result in different values if you run this personally on your machine!).

```
date_sql = """
SELECT
  end,
  datetime('now'),
  julianday('now') - julianday(end) AS days_old
FROM dt;
"""

# showing time since now
df = pd.read_sql(date_sql,con)
print(df)
```

```
                   end      datetime('now')    days_old
0  2024-01-01 12:40:00  2024-06-23 18:09:37  174.228911
1  2024-01-03 16:00:00  2024-06-23 18:09:37  172.090022
2  2024-01-04 12:40:00  2024-06-23 18:09:37  171.228911
3  2024-01-05 10:00:00  2024-06-23 18:09:37  170.340022
```

For CompStat like reports, we often want to calculate aggregate crime trends over time. To do this, we may create aggregated values for each day. Here is an example of counting up crimes per day based on the begin date. (As an analyst note, for these reports you would often use the *report* date, not the begin/end crime committed times.)

```
date_sql = """
SELECT
  SUM(CASE WHEN crime = 'Robbery' THEN 1 ELSE 0 END)
    AS Robbery,
  SUM(CASE WHEN crime = 'Burglary' THEN 1 ELSE 0 END)
    AS Burglary,
  SUBSTR(begin,1,10) AS bd
FROM dt
GROUP BY bd;
"""

# This table will now have four rows
df = pd.read_sql(date_sql,con)
print(df)
```

```
   Robbery  Burglary          bd
0        1         0  2024-01-01
1        0         2  2024-01-03
2        1         0  2024-01-04
```

Here I use `CASE WHEN` inside of an aggregation function, `SUM` to get multiple counts of specific crime types over time. You could also do `GROUP BY bd, crime`, but this will return the dataset in a different format. I often find this `CASE WHEN` and summing different 0/1 data values convenient in different types of data aggregation schemes.

> Note
>
> If you wanted to aggregate crime counts per month, one way would be to strip out the year and month fields, and group by them, e.g. `GROUP BY SUBSTR(begin,1,4), SUBSTR(begin,6,2)` would get the year and month respectively for this example. Some databases have functions to extract different date and time elements directly from a datetime variable.

Final point I want to make in this section is that the above table *only returns rows that are in the data*. So the day `2024-01-02` is missing from the result set. If you drew a line graph of crimes per day over time, you need to include 0 rows in the data for the chart to look correct. This is a scenario where using python in conjunction with SQL can be quite handy – it is often easier to fill in the missing dates in python than SQL.

```
# in python creating filled in dates
date_range = pd.date_range(df['bd'].min(),
                           df['bd'].max())
date_range = pd.DataFrame(date_range.astype(str),
                          columns=['bd'])
```

```
# now merging in the counts
date_range = pd.merge(date_range,df,how='left')
date_range.fillna(0,inplace=True) # replacing nulls with 0
print(date_range)
```

```
           bd  Robbery  Burglary
0  2024-01-01      1.0       0.0
1  2024-01-02      0.0       0.0
2  2024-01-03      0.0       2.0
3  2024-01-04      1.0       0.0
```

There is a way to do this directly in SQL – some databases have a function that you can generate a series of values. Here is an example of doing that in SQLite as an illustration.

```
date_sql = """
WITH RECURSIVE generate_series(value) AS (
  SELECT MIN(julianday(SUBSTR(begin,1,10)))
    FROM dt
  UNION ALL
  SELECT value+1 FROM generate_series
   WHERE value+1<= (SELECT MAX(julianday(SUBSTR(begin,1,10)))
    FROM dt)
), days AS (SELECT date(value,'julianday') AS bd
   FROM generate_series),
   aggv AS (SELECT
  SUM(CASE WHEN crime = 'Robbery' THEN 1 ELSE 0 END)
    AS Robbery,
  SUM(CASE WHEN crime = 'Burglary' THEN 1 ELSE 0 END)
    AS Burglary,
  SUBSTR(begin,1,10) AS bd
FROM dt
GROUP BY bd)

SELECT
  days.bd,
  COALESCE(aggv.Robbery,0) AS Robbery,
  COALESCE(aggv.Burglary,0) AS Burglary
FROM days
LEFT JOIN aggv
  ON aggv.bd = days.bd
"""

# This table will now have four rows
df = pd.read_sql(date_sql,con)
```

```
print(df)
```

```
           bd  Robbery  Burglary
0  2024-01-01        1         0
1  2024-01-02        0         0
2  2024-01-03        0         2
3  2024-01-04        1         0
```

This is a complicated query, so I leave it as an exercise to the reader to go and figure out how this query works (I don't tend to use `RECURSIVE` queries, but I do on occasion use the tools to generate series directly in different SQL databases). This also uses the SQL function `COALESCE`, which is one way to fill in null values. Here it replaces the null values with 0, which is what we want.

7.8 Connecting to Different Databases

The examples in this chapter have used an in-memory database, SQLite, as a simple way to demonstrate using SQL queries. In real life applications, when querying your data it will likely *not* be from a SQLite database. Different police record management systems have different databases, but a much more common database in my experience is Microsoft's SQL Server, web sites often use the MySQL database (an Oracle product), and a popular open source database is Postgres. Many crime analysts I know use Microsoft Access (it is possible to have a database in one version, and simply link it to another database application, this is commonly how people use Microsoft Access in my experience).

There are different ways to connect to these databases using python code. The two most common methods are via the `pyodbc` library, or via the `sqlalchemy` library. (If you are connecting to a cloud based database, it will likely be using `sqlalchemy`, whereas `pyodbc` is more common for Windows machines and local servers.) Here I will show different examples of connecting to different databases to help you get started. But you will need to figure out how to connect to the data on your local system to be able to run SQL code like I am showing in this chapter. Note these code samples are not run, as they would only work if you happened to have these specific databases on your personal computer.

```
# example using pyodbc connection
import pyodbc

# can see the available drivers
print(pyodbc.drivers())

# creating a connection string
user = 'Username'     # see next section on
pwd = 'Yourpassword' # how to keep this secret
serv = 'ServerName'  # and not have your password
db = 'DatabaseName'  # saved in the code
driv = 'ODBC Driver 17 for SQL Server'
```

```
cstr = (f'DRIVER={drv};DATABASE={db};SERVER={serv};'
        f'UID={user};PWD={pwd}')

# now connecting
con = pyodbc.connect(cstr)

# this is same as sqlite con earlier
pd.read_sql('SELECT * FROM database.table_name LIMIT 5',con)
```

After you install the `pyodbc` python library. If you import the library and run `pyodbc.drivers()`, you can see the OBDC drivers you have installed on your Windows operating system. This is again very idiosyncratic to your local system – for example here I have as the driver version 17 for SQL Server, but on your system you may have a different version installed (or need to install a different version to connect directly to your particular database). If you google your database name plus pyodbc, there is a good chance you will be able to find examples online of constructing the correct pyodbc connection string. The *server* location in the above string is then the location on your saved drive the database can be accessed from.

Many databases should out of the box have ODBC drivers to install for analysts to use, so often the bigger hurdle is figuring out the correct server string to use. Note this presumes that you have direct access to the database on your system, most of the time available via the *intranet* for most police departments.

A second popular python library to connect to databases is via the `sqlalchemy` library. Here is an example of creating a connection string to query a Postgres database in python. You don't technically need to import the `psycopg2` library in your code, but it is necessary for the library to be installed to be able to query a Postgres database using sqlalchemy.

```
# example using sqlalchemy
import sqlalchemy
import psycopg2

uid = 'postgres' # in real life, read these in as secrets
pwd = 'mypass'   # or via config files
ip = 'server.location:5432' # address and port
conn_str = f'postgresql://{uid}:{pwd}@{ip}/postgres'
eng = sqlalchemy.create_engine(conn_str)

# this is same as sqlite con earlier
pd.read_sql('SELECT * FROM database.table_name LIMIT 5',eng)
```

You can see that this is similar in nature to pyodbc – you simply need to create a string formatted in a particular way, then create a connection using that string. Note here that the ending `/postgres` is the default postgres schema, so you may need to swap out a different schema name for your actual database.

> Note
>
> If your password has special characters in it, you may need to url encode the password. For example, you would include `from urllib.parse import quote` in your import statements, and then use `pwd = quote('pass@withspec!alchars')`.

One last not, when connecting to different databases, sometimes instead of having the connection object as a constant parameter in the script, you want to use a *context* manager. This would look like:

```
cstr = '...connection string...'
with pyodbc.connect(cstr) as con:
    table = pd.read_sql('..SQL...',con)

# then do manipulations with the table
table.plot(....)
```

So this format does not return a connection object, it is only an active connecting during the with statement. You cannot then use `pd.read_sql` except for within the `with` statement. The reason for this is that some databases (including SQLite) you only want to have a connection for a limited time, you do not want to leave the connection to the database during the rest of the script.

Imagine you had a script that once it reads the data, does more intensive analyses (like a machine learning model). Your connection object may block other queries for some databases, which is not good practice. Especially if you are querying from a database that multiple people need to use.

> Note
>
> This is also a common way to work with external files in python. Say you wanted to read a file or append to a text file. That would look like:
>
> ```
> # This reads line from a file
> with open("file.txt", "r") as file:
>     res = file.readlines()
>
> # res will be a list with each
> # line of the file a separate
> # element
>
> # This adds lines to a file
> with open("file.txt", "w") as file:
>     file.write("\nAdding in a new line")
> ```
>
> Using the context manager prevents keeping the file being locked to view and manipulate by other processes.

7.9 Setting Secrets via Environment Variables

If your database requires a username and password, you should not include those directly in your scripts. That is, you should not have a line in your `.py` script that looks like:

```
pwd = 'MySupersecret101Password!'
```

If you share the script, someone else will be able to access your password. Such information software engineers call "secrets". They are things that someone else should not be able to see in your code.

The most common way to solve this is via having *environment variables* contain user sensitive information. You can access environment variables in python via the `os` library.

```
import os

# getting environment variables
user = os.environ.get('DB_USERNAME')
pwd = os.environ.get('DB_PASSWORD')
```

I will show how to set these secret variables on the Windows operating system. It is possible to set them though for all operating systems. In Windows, to set environment variables, search for "env" in the search bar in the lower left. A dialogue to edit the system environment variables will pop up.

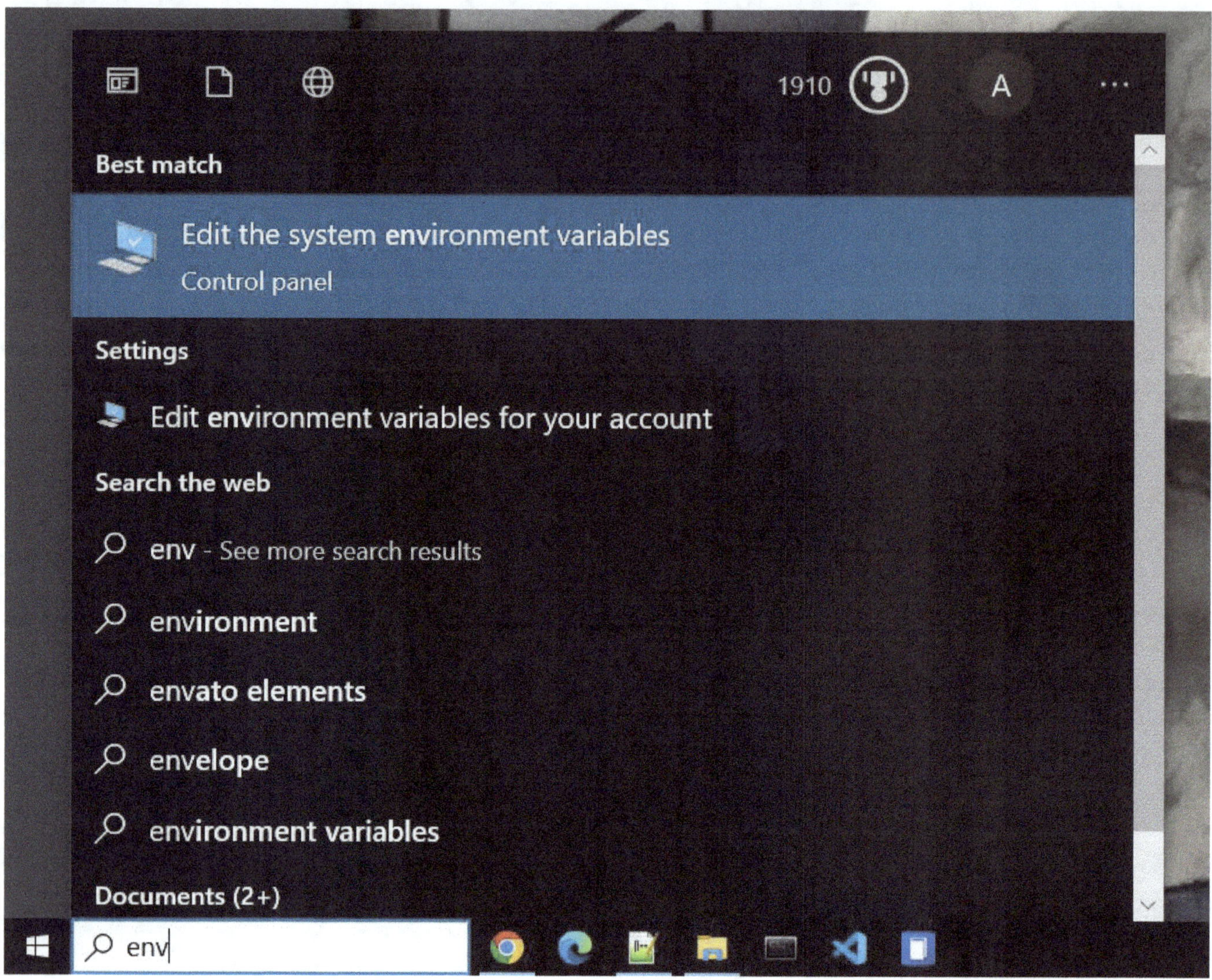

If you click that result, you then get a page where you can select the button `Environment Variables`.

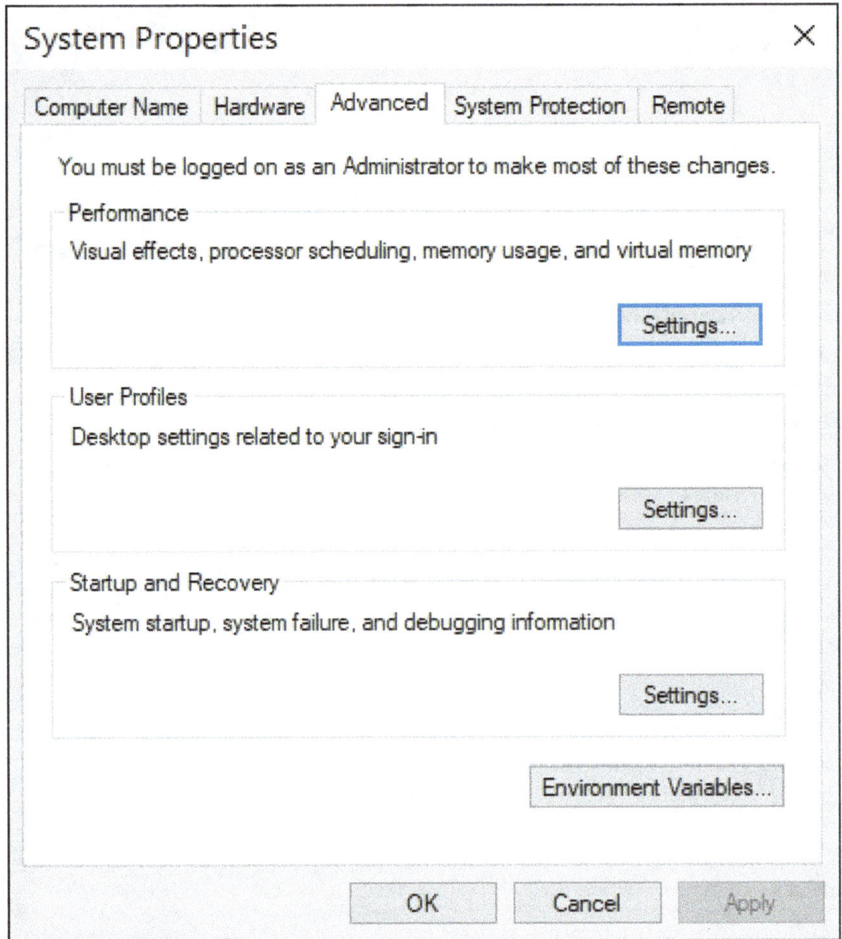

Then a final page where you can edit/add new environment variables. In my above example, my code expects the environment variable to be named as `DB_PASSWORD`, and in the variable value you would place your password.

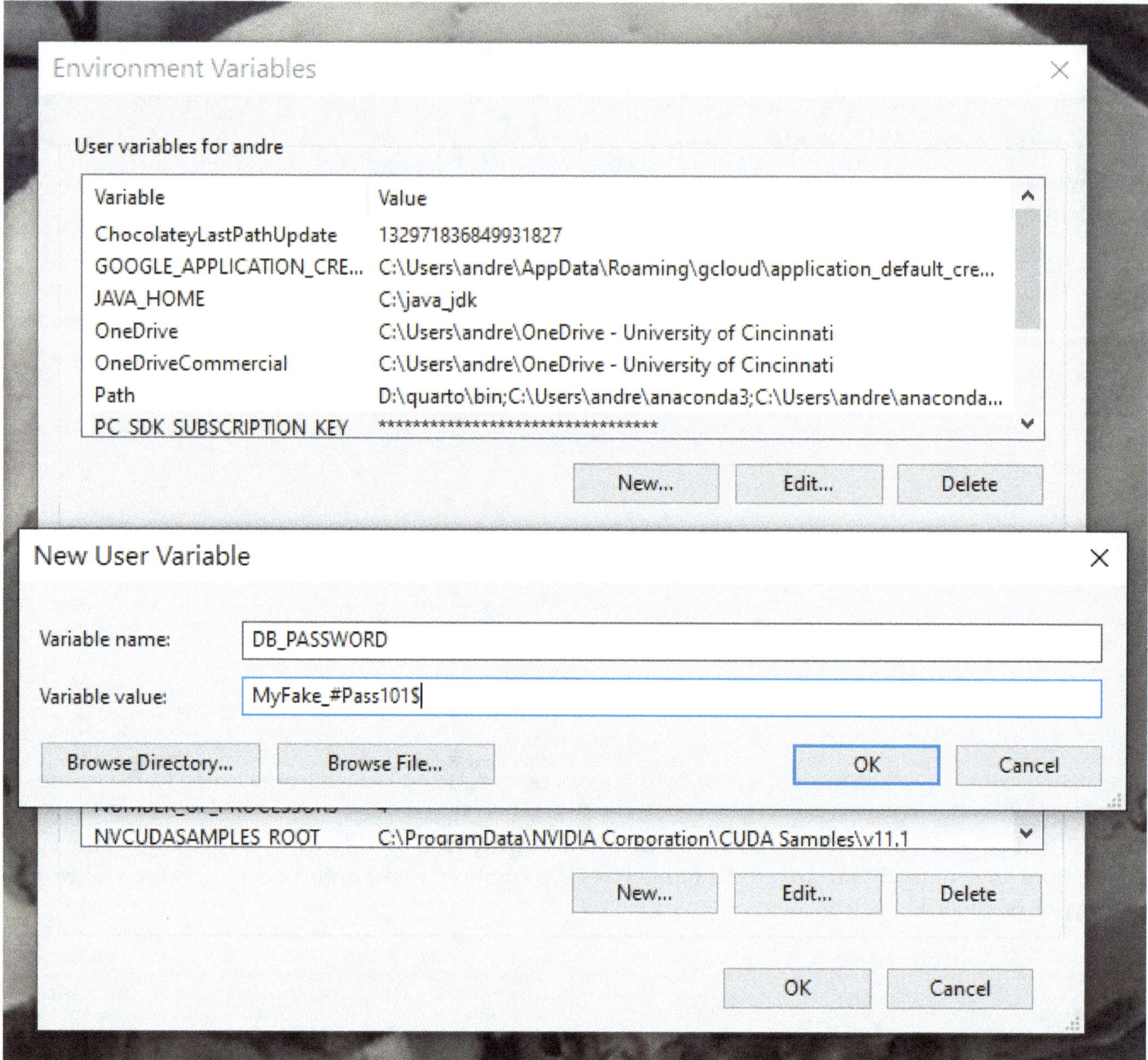

For those who use Mac's or Unix, you can set the environment variables via your bash profile.

> Note
>
> An alternative to setting secrets is to create a file that contains this information, and then load that variable in your code. This approach works if you are sharing code via github (just make sure to not share the file with your secrets!). But does not if you are writing code in a shared drive.
>
> For either approach, you should place a `README.md` file at the root of your project that details all of the environment variables (or config files) that need to be set to be able to run the code.

7.10 More Advanced SQL Examples

So far, I have shown basic examples of generating SQL queries, creating connection strings, and setting up environmental variables in python code that contains sensitive information like passwords. The final section consists of several more examples of data manipulation I have personally found useful in my career when using SQL. All of the examples so far haven't touched on *why* one would use some of the SQL tricks I have shown, such as creating crime counts per day, over doing the same operations by reading in a table directly into a pandas dataframe and doing the operations on the original data table.

All of the operations you can do in SQL you can do in pandas, and the majority of operations *you would want to do* you can do directly via SQL. SQL is a convenient way *to share code* however. For one example, you can *create a view*. A view is just a way to store a piece of SQL code that returns a table so others can use it. Here is an example of calculating year-to-date statistics, comparing this year to last year.

To start though, often when you are querying a database, data analysts do not have what is called *write* access to the data. You often *can only* read data. All of the operations shown so far are OK for reading, *except* for when I created a table and inserted values. When you have write access, it is very important to not make any breaking changes to a database. Often times analysts work on a *copy* of the production database, in which they have more control over. And if errors occur it is not a big deal (since it is just a copy). For example, I commonly have python code that copies from a production database, and saves the new values to a special analyst database.

```
table_names = ['Crime','Charges','Arrests']
for t in table_names:
    # read from production database
    df = pd.read_sql(t,prod_conn)
    # write to analyst database you have control over
    df.to_sql(t,analyst_conn,index=False)
```

And then that code runs once a night on a schedule (when you are not at work), to update your local analytical database. Note that it has two different connections, a `prod_conn` to read from the production database, and a `analyst_conn` to write to a different database.

> Note
>
> Another approach that is common is having *a linked database*. So you link one database to the other tables. For example you can have one type of database, but have Microsoft Access link to that databases tables, and then write queries to Access that are passed through to the original database.

This process of extracting data from one location and loading it into another location is commonly referred to as ETL, extract-transfer-load. You often don't want to be running data intensive queries on the production record management system (or other production database), which is meant for many small reads and writes, such as when a report is filed or someone calls 911. (In these examples, when you establish a connection to a sqlite database, *only you* can make reads or writes. Most databases allow multiple connections, but the more connections are currently being made typically slow down the overall database, as well as when you do large queries.) Doing an ETL job once a night, and then running your queries against an analyst database is a way to make sure you don't mess up the production database unintentionally.

> Note
>
> If the tables are very large, to prevent running out of memory this workflow may look like:
>
> ```
> for t in table_names:
>     # read from production database in chunks
>     # so do not run out of memory
>     chunk_df = pd.read_sql(t,prod_conn,chunksize=10000)
>     # clear out old table
>     analyst_conn.execute(f'DELECT FROM {t};')
>     # write to analyst database you have control over
>     for ch in chunk_df:
>       ch.to_sql(t,analyst_conn,index=False,
>           if_exists='append')
> ```
>
> Reading the data in smaller chunks will prevent your local python environment from running out of RAM.

So here I show additional examples of more advanced SQL, but these should almost always be run against the analytical copy of the database, not the actual production database in which your current data is stored. First I show an example of adding a few dates to our prior table. This time I do it via adding the data directly in a pandas dataframe. Using `df.to_sql` by default *replaces* the table entirely. Specifying the `if_exists='append'` kwarg makes it so it adds rows to the table.

```
# Add in a few crimes for the previous year
id = [5,6]
beg = ['2023-01-07 12:30:00','2023-01-08 00:30:00']
end = ['2023-01-07 14:30:00','2023-01-08 00:50:00']
cr = ['Burglary','Robbery']
last_year = pd.DataFrame(zip(id,beg,end,cr),
            columns=['id','begin','end','crime'])
last_year.to_sql('dt',con,index=False,if_exists='append')

print(pd.read_sql('SELECT * FROM dt;',con))
```

```
   id                begin                  end     crime
0   1  2024-01-01 12:30:00  2024-01-01 12:40:00   Robbery
1   2  2024-01-03 08:00:00  2024-01-03 16:00:00  Burglary
2   3  2024-01-04 12:30:00  2024-01-04 12:40:00   Robbery
3   4  2024-01-03 08:00:00  2024-01-05 10:00:00  Burglary
4   5  2023-01-07 12:30:00  2023-01-07 14:30:00  Burglary
5   6  2023-01-08 00:30:00  2023-01-08 00:50:00   Robbery
```

Now we can make a query that calculates the year to date. So if today is 2024-02-23 (February 23rd in the year 2024), a common analysis is to compare the prior year during the same time period, 2023-01-01 through 2023-02-23. You can do this analysis directly in SQL, *for the current date*. Here I create what is called a view to save that SQL query in a standard way.

```
# Creating a view for YTD stats
view_ytd = """
CREATE VIEW ytd_stats AS
WITH cy AS (
  SELECT
    crime,
    strftime('%Y-%m-%d','now') AS CurrDate,
    COUNT(*) AS CurrYear
  FROM dt
  /* selecting the current year of data
     note if you run this in future years, it wont quite work
     in these tables */
  WHERE
    strftime('%Y','now') = strftime('%Y',begin)
  GROUP BY crime
), ly AS (
  SELECT
    crime,
    COUNT(*) AS LagYear
  FROM dt
  /* this gets the prior year */
  WHERE
    CAST(strftime('%Y',begin) AS int) =
   (CAST(datetime('now','-1 Year') AS int))
  GROUP BY crime
)

SELECT
  cy.*,
  ly.LagYear
FROM cy
LEFT JOIN ly
  ON cy.crime = ly.crime;
"""

# This does not return a table directly
# just registers the view
con.execute(view_ytd)
```

```
<sqlite3.Cursor at 0x217da2eb7c0>
```

Now it is very simple to call this view, it is simply `SELECT * FROM ytd_stats;`.

```
# Now can query the simpler view
print(pd.read_sql('SELECT * FROM ytd_stats;',con))

print('\nYou can treat this like any table\n')
query = "SELECT * FROM ytd_stats WHERE crime = 'Burglary'"
print(pd.read_sql(query,con))
```

```
      crime    CurrDate  CurrYear  LagYear
0  Burglary  2024-06-23         2        1
1   Robbery  2024-06-23         2        1

You can treat this like any table

      crime    CurrDate  CurrYear  LagYear
0  Burglary  2024-06-23         2        1
```

Note

Many people (including myself) have criticized this approach, especially when calculating percent change stats year over year. See https://andrewpwheeler.com/2023/03/25/a-statistical-perspective-on-year-to-date-metrics/ for discussion. I include it as it is a good illustration of how knowing SQL can simplify different repeated data analyses.

Now *everyone* who has access to this database can write that simpler query, and get year to date stats, all defined the same way, up to date. It is a very nice tool to ensure consistency across data products for multiple people.

Note

Another approach to do the YTD stats is instead of getting the current date directly in SQL via code such as `strftime('%Y-%m-%d','now') AS CurrDate`, you can use string formatting and insert the date into the query string. See the prior chapter on strings for examples of this. For example, you could create a query using f-strings such as `query = f"... WHERE date < {end} AND date >= {begin}"`, where `end` and `begin` are python string objects that have the dates you want to fill in the query.

Sometimes when you have views, you want to update them or drop them. You can drop entire tables or views via `DROP TABLE tablename;` or `DROP VIEW viewname;`. Here I also show a SQL command to see the current tables.

```
# this SQL shows the current tables
print(pd.read_sql('SELECT * FROM sqlite_schema',con))
```

```
# drop the dt table and the view
con.execute('DROP TABLE dt;')
con.execute('DROP VIEW ytd_stats;')

# now these are gone from table
query = 'SELECT type,tbl_name FROM sqlite_schema'
print(pd.read_sql(query,con))
```

```
    type       name   tbl_name  rootpage  \
0  table    Officer    Officer         2
1  table      Crime      Crime       320
2  table         dt         dt       441
3   view  ytd_stats  ytd_stats         0

                                                 sql
0  CREATE TABLE "Officer" (\n"ori" TEXT,\n  "ncic...
1  CREATE TABLE "Crime" (\n"ori" TEXT,\n  "aggrav...
2  CREATE TABLE dt (\n  id int,\n  begin text,\n ...
3  CREATE VIEW ytd_stats AS\nWITH cy AS (\n  SELE...
    type tbl_name
0  table  Officer
1  table    Crime
```

The command `SELECT * FROM sqlite_schema` changes from database to database, but there are often SQL queries you can use to get *meta-data* for the database, such as table names.

If instead you want to just delete specific rows from a table, you can use

```
DELETE FROM table WHERE ...
```

Omitting the `WHERE` statement, e.g. `DELETE FROM tablename;`, just deletes all rows in the entire table (but the table still exists).

One SQL concept I have not covered is *indexing*. Indexes are typically defined as fields that are unique identifiers for a table. This is common in introductory SQL texts, as indexing certain columns on your database can improve look up times. These however are almost always to improve lookup times for individual rows. Using indexes does not improve large analytic queries as are common in data science applications. So I do not cover them here, but are important to know about if you plan on pursuing a career more oriented towards data management.

There is one area though that it is useful for data analytics, in particular for ETL jobs, you can use what is called an *UPSERT* to add in new data to a table. Instead of simply appending, if you upload a data field for an updated record, it will overwrite the prior value. Here I am going to make a new table to illustrate, based on ORIs and violent crime in the other tables.

```
# creating a smaller table
crime_table = """
CREATE TABLE ct (
  ori text PRIMARY KEY,
  viol int,
  pop int,
  name text
);"""

# creating new table
con.execute(crime_table)

# can show column names and types
print(pd.read_sql('PRAGMA table_info(ct);',con))
```

```
   cid  name  type  notnull dflt_value  pk
0    0   ori  TEXT        0       None   1
1    1  viol   INT        0       None   0
2    2   pop   INT        0       None   0
3    3  name  TEXT        0       None   0
```

Now we can insert a small selection of rows into this table, again using `INSERT`, but this time as defined via a different query.

```
# creating a smaller table
crime_insert = """
INSERT INTO ct (ori,viol,pop,name)
SELECT
  Crime.ori,
  Crime.violent_crime,
  Officer.population,
  SUBSTRING(Officer.ncic_agency_name,1,10)
FROM Crime
LEFT JOIN Officer
  ON Crime.ori = Officer.ori
WHERE Officer.population > 1000000
"""

# inserting rows from other tables
con.execute(crime_insert)

# can see the rows added now
print(pd.read_sql('SELECT * FROM ct;',con))
```

```
          ori   viol      pop        name
0   ILCPD0000  14321  2652124  CHICAGO PD
1   TXDPD0000  10009  1286121  DALLAS POL
2   NV0020100   8605  1667961  LAS VEGAS
3   AZ0072300  13515  1637902  PHOENIX PO
4   MD0160400   1494  1030232  MONTGOMERY
5   PAPEP0000  16202  1555812  PHILADELPH
6   VA0290100   1212  1094682  FAIRFAX CO
7   CA0371100   5932  1377838  SAN DIEGO
8   TXHPD0000  25987  2276533  HOUSTON PO
9   TX1010000   9199  2029908  HARRIS CO
10  TXSPD0000  12935  1465608  PD SAN ANT
11  NY0303000  61293  8236567  NEW YORK C
12  NY0510100   1201  1355324  SUFFOLK CO
13  NY0290000   1748  1087353  NASSAU COU
14  CA0194200  31772  3809182  LOS ANGELE
```

So here instead of adding in rows of data via the `INSERT ... VALUES (...)`, you can insert values from one table into another table using a select statement. Say we then wanted to change the Chicago PDs numbers, here lets change population to 2 million and violent crime to 14000.

```
# updating Chicago's stats
crime_update = """
INSERT INTO ct (ori,viol,pop)
VALUES ("ILCPD0000",14000,2000000)
ON CONFLICT(ori) DO UPDATE SET
  viol = excluded.viol,
  pop = excluded.pop
"""

# inserting the updated data
con.execute(crime_update)

# can see the rows added now
query = "SELECT * FROM ct WHERE ori = 'ILCPD0000';"
print(pd.read_sql(query,con).T)
```

```
                0
ori     ILCPD0000
viol        14000
pop       2000000
name   CHICAGO PD
```

Some databases have an explicit `UPSERT` command, but this is one way to change data for specific rows in a database that have defined primary keys. The prior ETL example I gave with reading from a production database and writing those tables to an analytics database could be made even more efficient by only querying recently updated values and using an upsert query to change or append those new values, instead of dropping and recreating the entire large table.

> Note
>
> A common approach to doing bulk updates like this is to store a table in a staging table, and then have a *stored procedure* append that temporary table to the main datatable on a regular cadence. So in python you could do `data.to_sql('load_table',con)`, and then run in SQL `INSERT INTO prod_table .... SELECT * FROM temp_table`.

Similar to the numpy and pandas chapter, I have only scratched the surface of what is possible with SQL and manipulating data in databases. There are ultimately *many* different aspects of SQL that I have not covered. Understanding SQL is not just good for crime analysts, but there are whole jobs devoted to the task. So learning SQL well will not only make you a better data analyst, it will also make you more marketable for a wide variety of roles.

While these prior two chapters showed working with tabular structures, you would not include a table with 1,000s of rows in a report for the police department. You would summarize that data, often in graphical form, for your reports. The following chapter shows how to create graphs in python using the matplotlib library.

8 Making Graphs with matplotlib

There are many different plotting libraries in python, but the main one I am going to focus on in this chapter is the *matplotlib* library. matplotlib is mostly used to create static graphs. That is, you have a set of data and you create a bar chart or line chart of that data, and then save it to a png or some other type of static file. You may then include that static graph in some other report, or use it for just yourself for exploratory data analysis. They are not interactive (like a dashboard).

This may seem limited, but many of the other plotting libraries in python use matplotlib under the hood. So I believe understanding the basic matplotlib library is a good place to start, as opposed to using a different library.

> Note
>
> Several of the alternative libraries to check out after you have finished this chapter are *seaborn* or *plotnine*. Both have conventions to make small multiple plots or functional aggregations similar to the ggplot R library. Each of these libraries use matplotlib under the hood, hence why I use matplotlib in this chapter. For interactive graphics, Vega-Altair or Plotly are currently two popular libraries. There are many competitors and changes in interactive graphics libraries though, so this advice is likely to go out of date quickly.

8.1 Creating a simple graph and saving the file

This first example shows off creating a very simple line graph. Similar to how `import pandas as pd` is a very common code convention, for matplotlib `import matplotlib.pyplot as plt` is the typical import statement.

```
import pandas as pd
import numpy as np
import matplotlib.pyplot as plt

x = [1,2,3]
y = [0.5,1.2,0.8]

fig, ax = plt.subplots()
ax.plot(x,y)
fig.show()
```

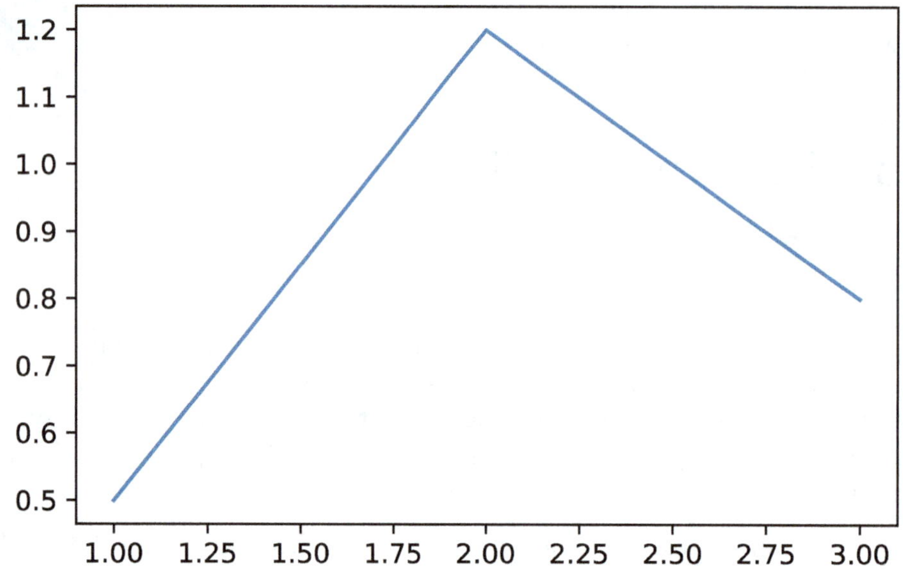

Plotting with matplotlib can seem confusing since it introduces many new types of objects. Here we first have `fig, ax = plt.subplots()`, this creates a figure and an axis object. Then we have `ax.plot(x,y)`, which adds a line to the `ax` object. Here I use a simple list to add in the numeric x and y values, but those could also be numpy arrays or pandas series objects as well. Finally we have `fig.show()`, that displays the image we built.

> Note
>
> Some uses of matplotlib use the `plt` library directly, e.g. use `plt.show()` or `plt.xticks` to set the tick marks for example. It is clearer in code to use the `fig` and `ax` objects directly in most circumstances.

How this displays the image is dependent on how you are running your code. If you are running your code in a REPL session, it pops up an image in a separate window. If you are running your code in a Jupyter notebook (see the next chapter), it will display the plot directly in your document.

If you want to save an image directly to disk, you can use `fig.savefig('ImageName.png')`. This saves a png file of the plot. Here is an example with the prior plot, saving to a high resolution png file (a high *dots per inch*, DPI, ratio of 500), and no border (the `bbox_inches` argument).

```
x = [1,2,3]
y = [0.5,1.2,0.8]

fig, ax = plt.subplots()
ax.plot(x,y)
# This is not displayed in the book
# but will save it in the root directory
fig.savefig('LinePlot.png',dpi=500,bbox_inches='tight')
```

Most of the time for plots, I suggest to save as a png file between 100 and 500 DPI.

Note

Remember about filepaths, so this saves the image in the root folder wherever python is running from. If you had the code `fig.savefig('./images/ImageName.png')` it would save the image in a images folder that is below the current path. You could also explicitly set the full path, with something like `im_path = r'C:\path\folder\image.png'` and `fig.savefig(im_path)`.

For a few notes on image formats, there are *raster* images and *vector* images. Rasters are grids of colors, so if you zoom in far you can see the grid cells. These include image types of png or jpg. For raster images I typically suggest png, as it is *lossless*. jpg (or jpeg) tries to compress the images down. So they are smaller in size, but can have artifacts. This is good for pictures taken with your phone, but is not necessary for the types of statistical graphs I show in this chapter.

The graphs shown in this chapter are all vector representations, so you *could* also save directly as a vector format. Vector formats you can zoom in infinitely into the graph, and it will not lose its resolution.

Vector formats include pdf, svg, or Windows formats such as emf. In matplotlib for example, you could do `plt.savefig('LinePlot.pdf')`. These are not bad, but take slightly more care to make them interoperable depending on how the end user will view the graph. You can copy a png file easily into a different report or presentation, a pdf document can be more tricky. Also they can cause issues with fonts if you are not careful, or changes between different operating systems. For this reason, I suggest beginners just save graphs as high resolution png files.

8.2 Styling Plots

There are are many items that one manipulates in plots. But regular ones include the x and y axis labels, and a plot title. Each modification takes an additional line of code to modify the plot.

```
x = [1,2,3]
y = [0.5,1.2,0.8]

fig, ax = plt.subplots()
ax.plot(x,y)
ax.set_xlabel("X axis label")
ax.set_ylabel("Y axis label")
ax.set_title("I am a Title")
fig.show()
```

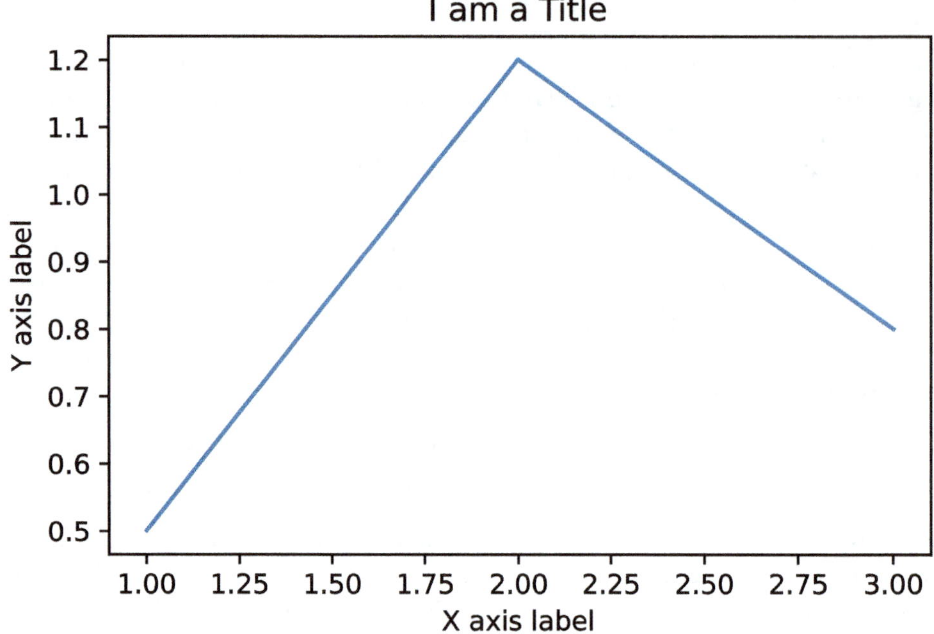

Note that the y axis label is vertical. If space allows, you can set it to be horizontal. Here I stretch out the plot size to be 6 by 3 inches as well in the `plt.subplots` call:

```
x = [1,2,3]
y = [0.5,1.2,0.8]

# adjusting the plot size to be 6 by 3
fig, ax = plt.subplots(figsize=(6,3))
ax.plot(x,y)
ax.set_xlabel("X axis label")
ax.set_ylabel("Y axis",rotation='horizontal',labelpad=20)
ax.set_title("I am stretched out a bit more")
fig.show()
```

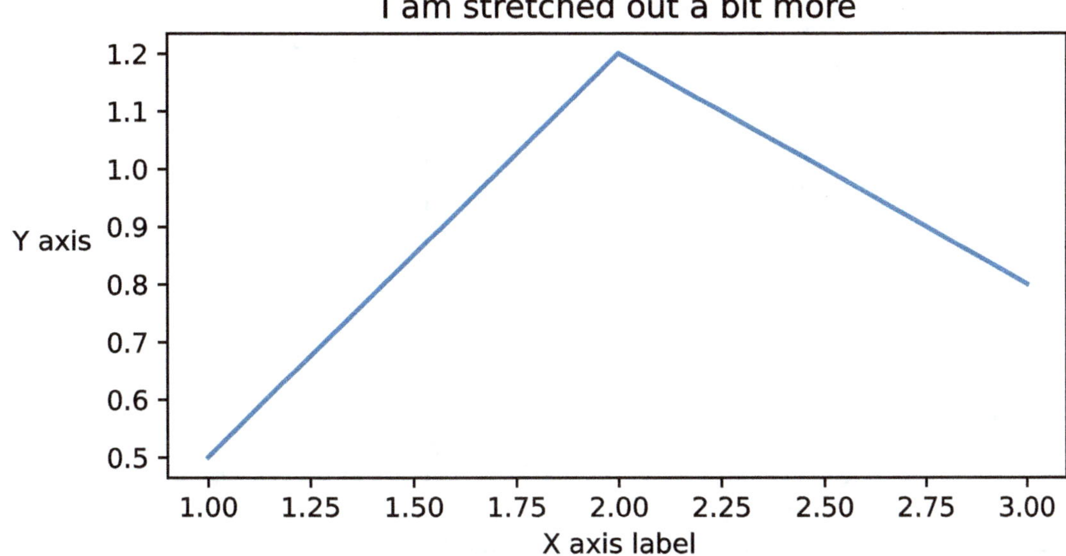

You need the `labelpad` argument here to make sure the text does not intrude on the y axis *ticks*. To adjust the ticks, you can use `ax.set_xticks` or `ax.set_yticks`.

```
x = [1,2,3]
y = [0.5,1.2,0.8]

fig, ax = plt.subplots()
ax.plot(x,y)
ax.set_xticks([1,2,3])
# can use numpy to make regular intervals
ax.set_yticks(np.arange(0.4,1.6,0.2))
ax.set_xlabel("X axis label")
ax.set_title("Title on the left",loc="left")
fig.show()
```

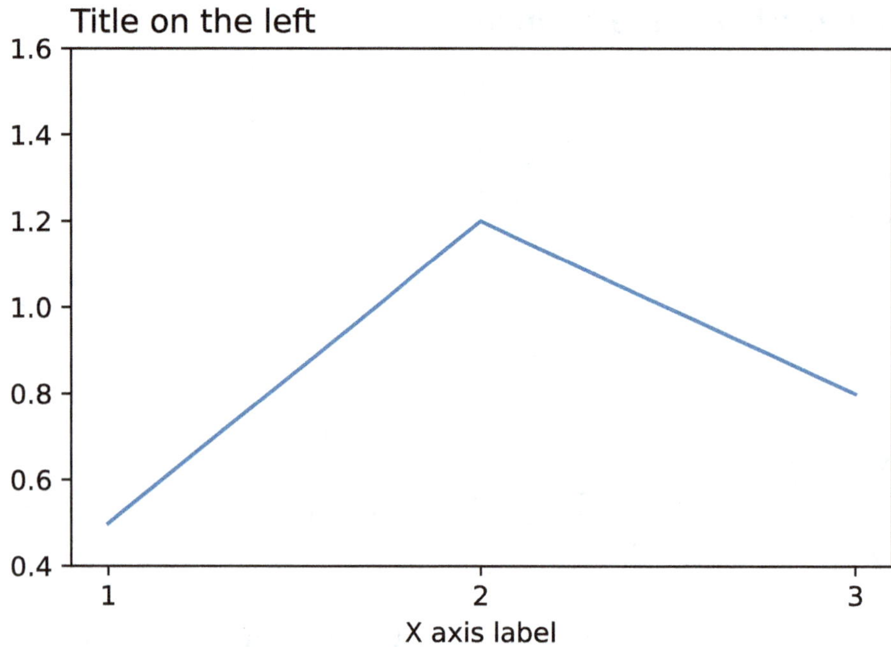

By default matplotlib tries to do smart tick label locations, but I commonly want to make my own. For one time graphs, it is good to tinker with aspects of graphs to make them look as nice as possible. In scenarios where you are auto-creating reports though (what I will go over more in the next chapter), you often need to rely on the defaults to a greater extent.

> Note
>
> When running code like `ax.plot(...)` or `ax.set_yticks(...)`, if running from the REPL you may get an output that is like `matplotlib.axis.Object` printed to the console. Similar to how if typing `x` from the REPL will print out the value that is associated with x. When automating reports, like I show next chapter, those will typically not be captured in the final output.

8.3 Plot design

There are many aspects of plots, but one distinction is between the formatting of the plot, the background color, gridlines, fonts, etc., vs the data the plot is actually showing. Pretty much everything for a graph is customizable, here I show a few examples of changing the font, background color, and gridlines for the graph.

```
x = [1,2,3]
y = [0.5,1.2,0.8]

# Ugly plot to showcase looks
fig, ax = plt.subplots()
```

```
ax.plot(x,y)
# Grey hexcode for background color
ax.set_facecolor("#DDDDDD")
# grid lines
ax.xaxis.grid(True)
ax.yaxis.grid(True)
# xlabel fontsize smaller
ax.set_xlabel("X axis label", fontsize=10)
# ytick different color
ylocs = [0.5,0.9,1.54]
ax.set_yticks(ylocs,labels=ylocs,color='red')
# Bold for title
ax.set_title("Bold Title",fontweight='bold')
fig.show()
```

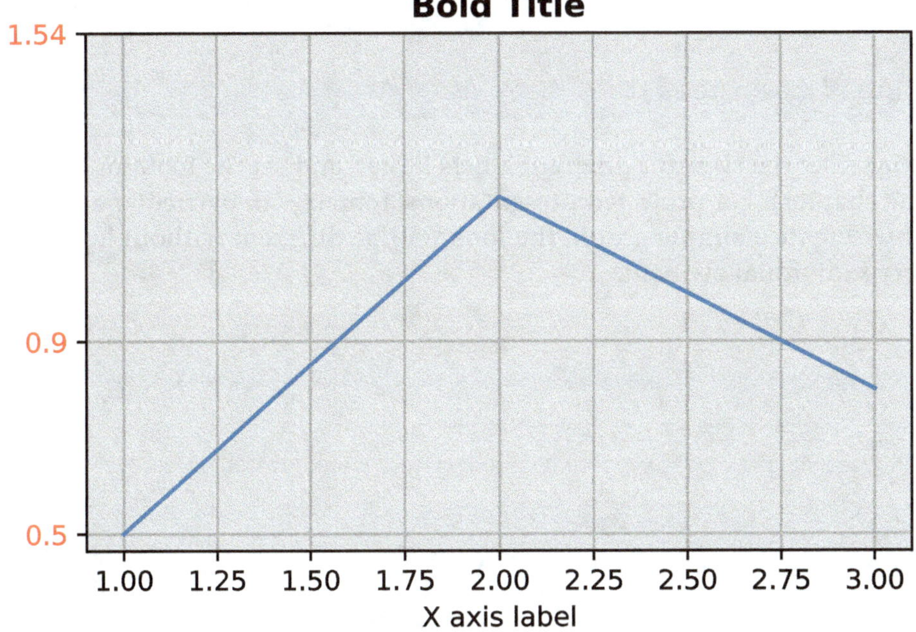

This is a very ugly graph! But hopefully illustrates the point that many of the graph options can be manipulated. This can be quite tedious though if you have many graphs, and you want to consistently change the looks all the time. To make consistent changes across all graphs, I like to create my own *template*, and apply that globally to all graphs. Here is my template for example.

```
import matplotlib

# my theme for graphs
new_theme = {'axes.grid': True,
             'axes.axisbelow': True,
```

```
              'grid.linestyle': '--',
              'legend.framealpha': 1,
              'legend.facecolor': 'white',
              'legend.shadow': True,
              'legend.fontsize': 14,
              'legend.title_fontsize': 16,
              'xtick.labelsize': 14,
              'ytick.labelsize': 14,
              'axes.labelsize': 16,
              'axes.titlesize': 20,
              'axes.titlelocation': 'left',
              'figure.dpi': 100}

# can print this to see all of the different
# options, it is a dictionary
#print(matplotlib.rcParams)

# updating the default
matplotlib.rcParams.update(new_theme)
```

Here I set by default dashed gridlines, set the sizes for different labels larger, and set the look for legends (which I will show a bit later in the chapter). For projected presentations, fontsizes in particular are often way too small by default. Now if we create a simple graph, the look will be different without having to type in all the extra looks for every individual element.

```
x = [1,2,3]
y = [0.5,1.2,0.8]

fig, ax = plt.subplots()
ax.plot(x,y)
ax.set_xlabel("X label")
ax.set_title("Title")
fig.show()
```

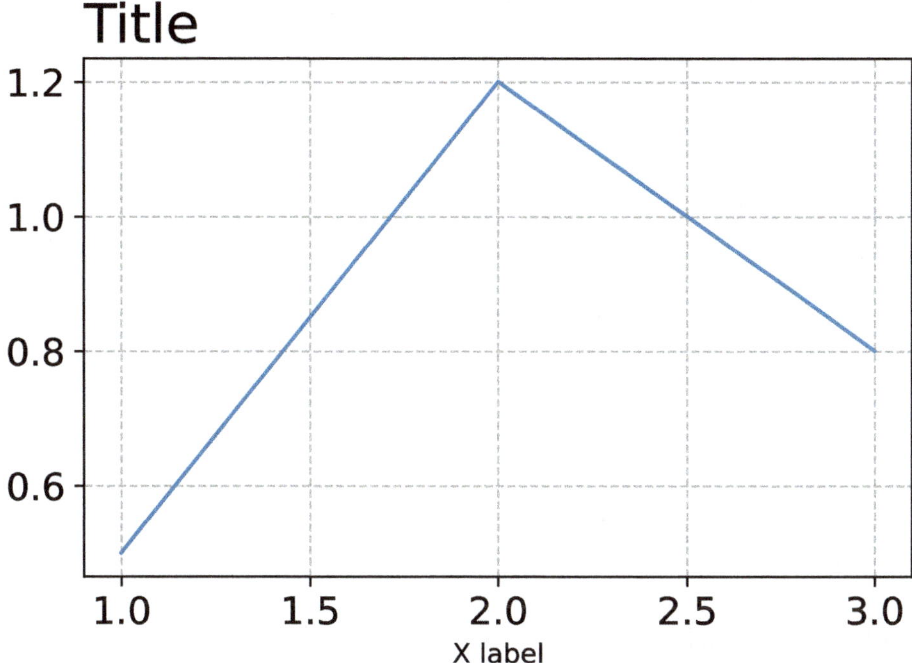

If you check out `print(matplotlib.rcParams)` you can see all of the different parameters you can tinker with in your plots.

> Note
>
> One thing you can change are the default colors used in graphs. You can either make your own, or use common color scales. My personal favorite to suggest is to check out https://colorbrewer2.org/ and their suggested palettes. The site is oriented towards maps, but they look very nice in other types of graphs as well in my opinion.

8.4 Line plots

The vast majority of graphs I create are collections of lines, bars, and circles. These next few sections will show some tricks when making these different types of graphs. For line graphs, I often like placing points on top of the lines.

```
x = [1,2,3]
y = [0.5,1.2,0.8]

fig, ax = plt.subplots()
ax.plot(x,y,marker='o')
fig.show()
```

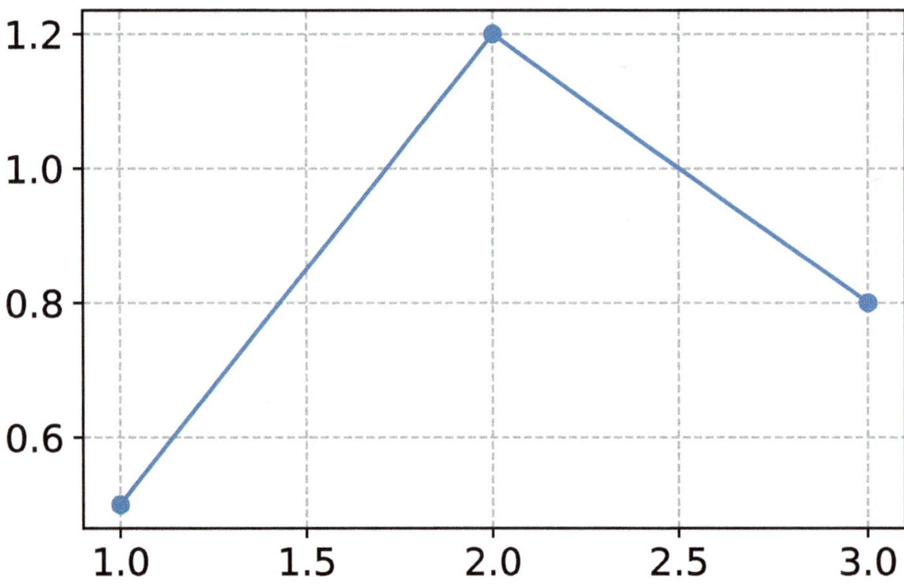

My favorite way to draw line graphs is to use white outlines for points. Here I show making a black, slightly thicker line.

```
x = [1,2,4]
y = [0.5,1.2,0.8]

fig, ax = plt.subplots()
# color='k' is shorthand for black
ax.plot(x,y,color='k',linewidth=1.5,
        marker='o',markeredgecolor='white',
        markersize=11)
fig.show()
```

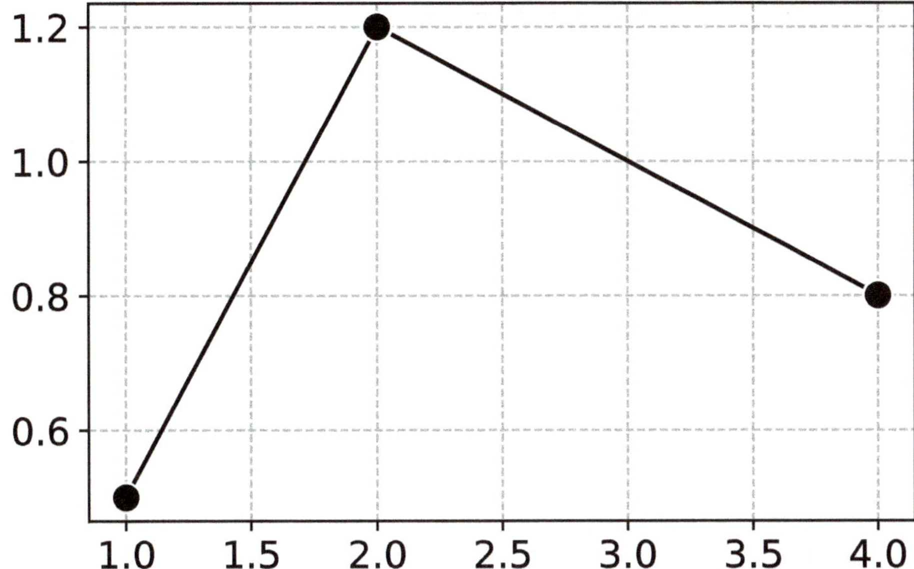

Note how above it is not a regular interval, it skips the x point at 3. Having the points superimposed on the plot clearly shows this. If you are plotting crime counts over time, you probably want to include zero values though in your plot. Here I show using a different color and marker type.

```
x = [1,2,3,4]
y = [0.5,1.2,0,0.8]

fig, ax = plt.subplots()
# this uses a hexcolor purple
ax.plot(x,y,color='#810f7c',linewidth=2,
        marker='x',markersize=12)
fig.show()
```

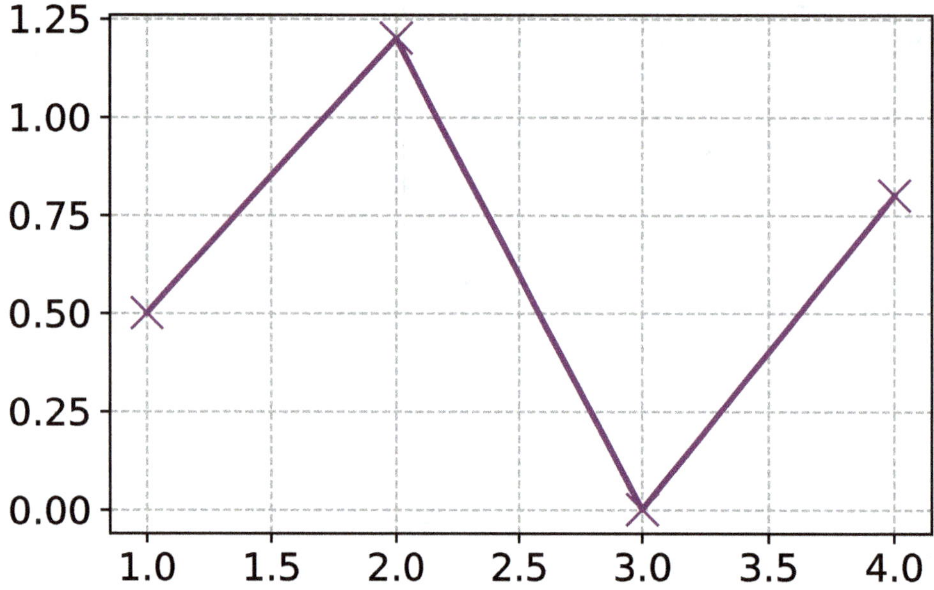

If you want to add in multiple lines, you would make multiple calls to `ax.plot`. This also shows what missing data looks like by default in line plots.

```
x = [1,2,3,4]

# missing data example
y1 = [0.7,0.1,None,1.0]
y2 = [0.5,1.2,0,0.8]

fig, ax = plt.subplots()
# two different lines
ax.plot(x,y1,color='k',linewidth=1.3,
        marker='o',markersize=11,
        markeredgecolor='white')
ax.plot(x,y2,color='#810f7c',linewidth=2)
fig.show()
```

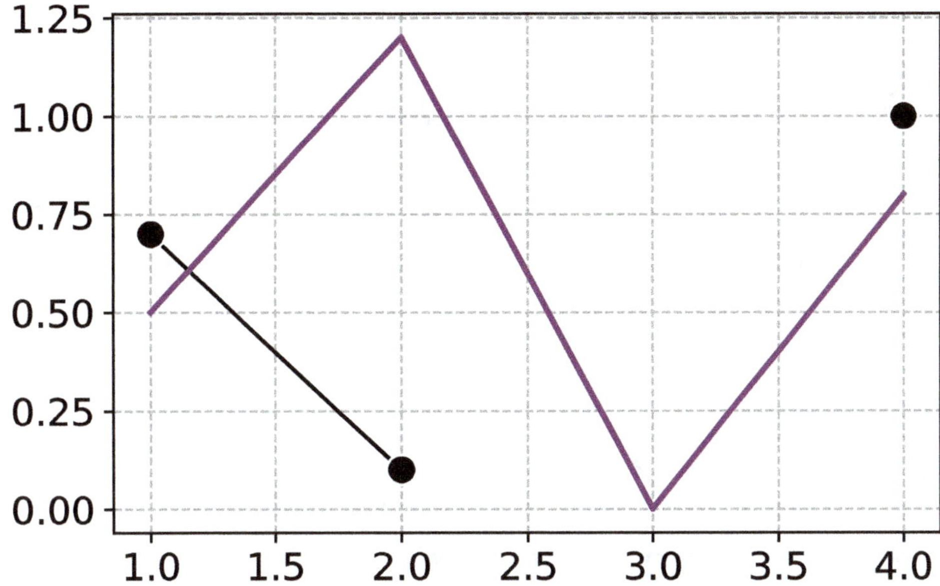

But now you have a problem, in that you don't know exactly what the black and purple lines represent. To distinguish between them, you can add labels to each line, and then place them in a legend.

```
x1 = [1,2,4] # no missing point
y1 = [0.7,0.1,1.0]
x2 = [1,2,3,4]
y2 = [0.5,1.2,0,0.8]

fig, ax = plt.subplots()
# two different lines
ax.plot(x1,y1,color='k',linewidth=1.3,
        marker='o',markersize=11,
        markeredgecolor='white',label='Y1')
ax.plot(x2,y2,color='#810f7c',linewidth=2,label='Y2')
ax.legend()
fig.show()
```

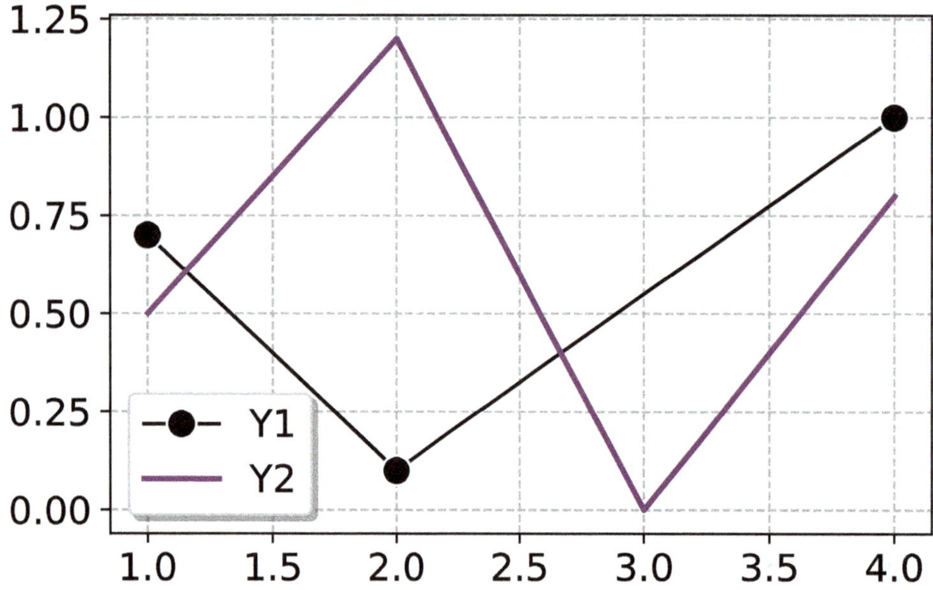

A few common legend manipulations I do are changing the location of the legend *inside* the plot, and changing the fontsize of the legend. Very busy plots it can be difficult to locate the legend so it does not cover up other items, and will show in a later section placing the legend outside of the plot area.

```
x1 = [1,2,4] # no missing point
y1 = [0.7,0.1,1.0]
x2 = [1,2,3,4]
y2 = [0.5,1.2,0,0.8]

fig, ax = plt.subplots()
# two different lines
ax.plot(x1,y1,color='k',linewidth=1.3,
        marker='o',markersize=11,
        markeredgecolor='white',label='Y1')
ax.plot(x2,y2,color='#810f7c',linewidth=2,label='Y2')
ax.legend(loc='upper left',fontsize=7)
fig.show()
```

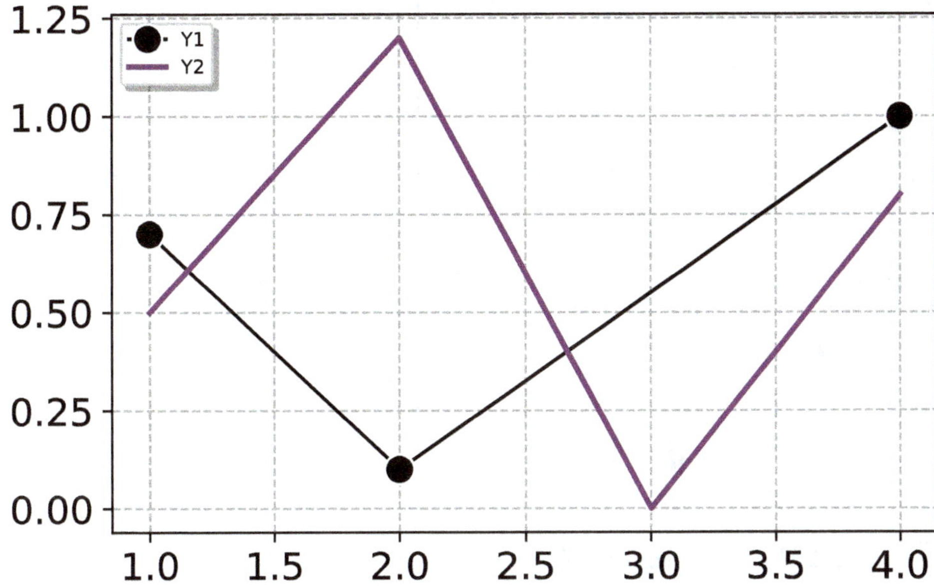

You can do `lower right`, `center left`, etc. to set the location. Matplotlib tries to figure out a location that does not cover up other plot elements by default.

Quite often line plots are displaying some metric over time. Here I do random crime counts per month over two years.

```
# random crime counts
np.random.seed(10)
x = pd.date_range('1/1/2023','6/1/2024',freq='MS')
y = np.random.binomial(20,0.5,x.shape)

fig, ax = plt.subplots()
# two different lines
ax.plot(x,y,color='k',linewidth=1.3,
        marker='o',markersize=11,
        markeredgecolor='white')
fig.show()
```

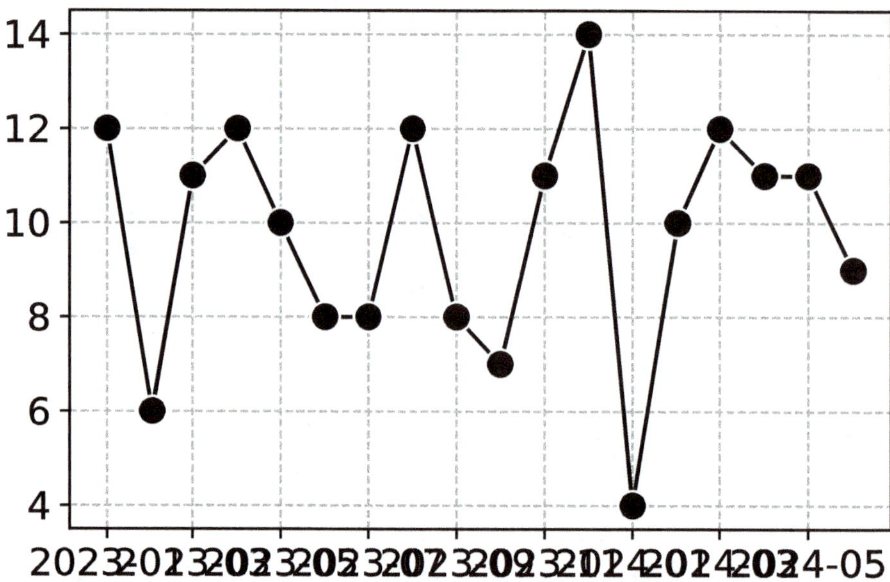

To make more room for the date labels, different things you can do are stretch out the plot horizontally, only display labels every quarter, make the font size for the labels smaller, and rotate the labels. Here are examples of each of these for this graph.

```
# wider plot
fig, ax = plt.subplots(figsize=(8,4))
# two different lines
ax.plot(x,y,color='k',linewidth=1.3,
        marker='o',markersize=11,
        markeredgecolor='white')
# set xticks every quarter
tr = pd.date_range(x.min(),x.max(),freq='QS')
# only month and year label
tr_labs = tr.strftime('%Y-%m')
ax.set_xticks(tr,tr_labs,rotation=30,fontsize=8)
fig.show()
```

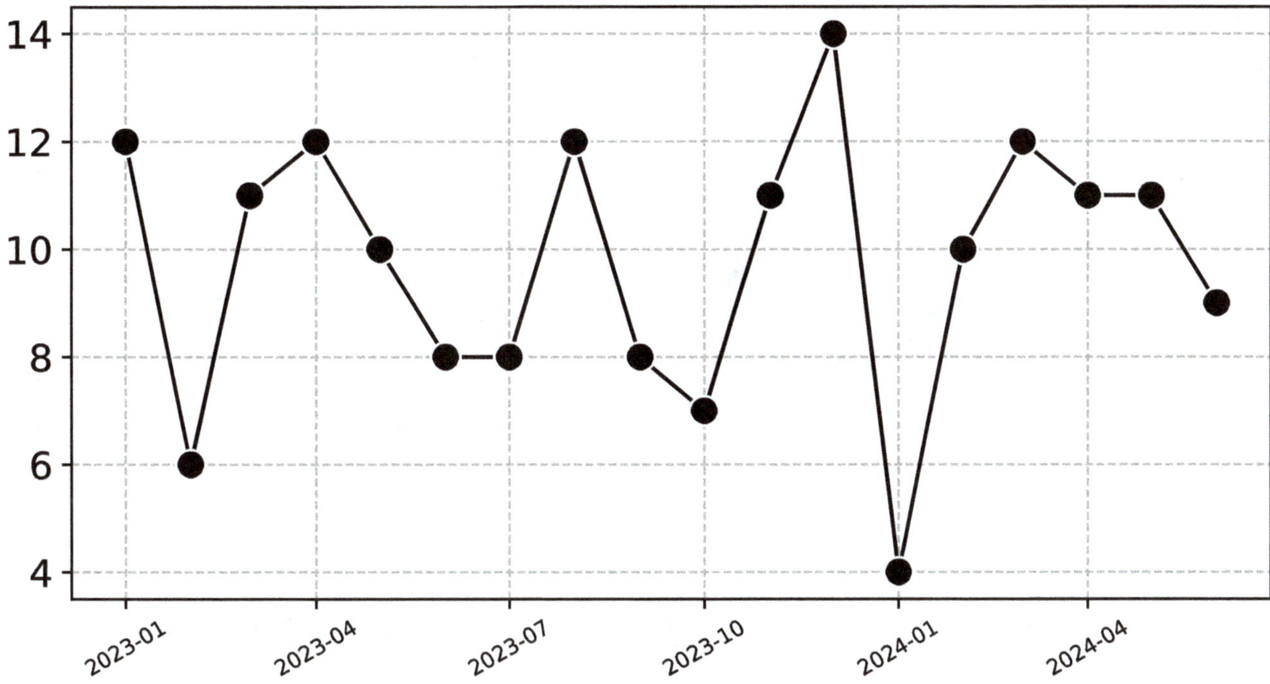

Another common element for line charts is to show an error range. Here I illustrate having a background area range on the plot, with superimposed data.

```
# calculating error bars as Poisson variance
low = y - 2*np.sqrt(y)
hig = y + 2*np.sqrt(y)

# superimposing error bars
fig, ax = plt.subplots(figsize=(8,4))
# use fill between
ax.fill_between(x, low, hig,
                alpha=0.2, color='k')
ax.plot(x,y,color='k',linewidth=1.3,
        marker='o',markersize=8,
        markeredgecolor='white')
ax.set_xticks(tr,tr_labs)
fig.show()
```

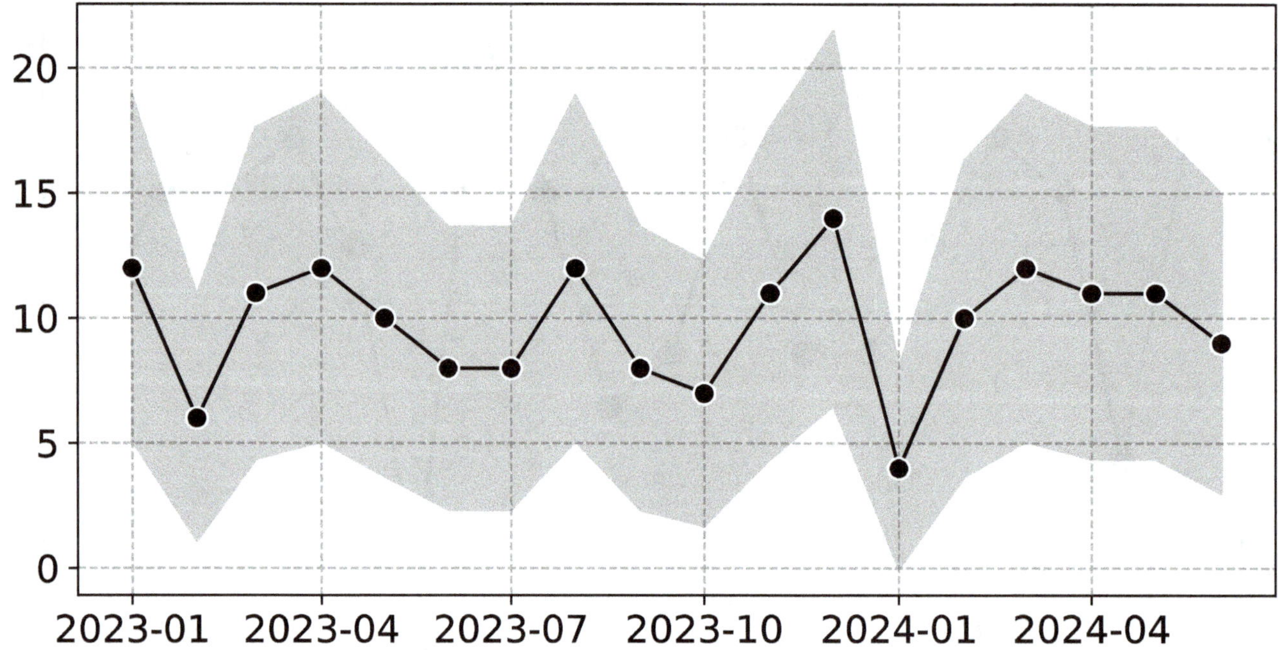

> Note
>
> Depending on what your template looks like and how you are rendering the images, your plot may look slightly different than the plot that I have in this book. I show all the different little tricks, such as making text smaller, rotating labels, or making fewer ticks, to help alleviate some of those issues. Often you need to experiment with a particular graph to get it to look exactly like you want, without elements colliding.

Here I show off the area as being semi-transparent, using the `alpha` argument. This allows you to see the gridlines below the error bar area. Even though I draw the error area as black, it shows up as lighter grey, since it is semi-transparent. Next chapter I will show how to do a similar chart to show when current crime counts are outside the normal range that you would expect.

A final example I want to show with line plots are vertical rules and shaded background areas. These are common to illustrate some policy event occurred, and you want to pay attention to the values before and the values after. Here you can do that via `ax.axvline`.

```
# making an abrupt change
y[-6:] = np.random.binomial(20,0.2,(6))

fig, ax = plt.subplots(figsize=(8,4))
# creating a vertical rule
ax.axvline(x[-6],color='blue',linewidth=2)
ax.plot(x,y,color='k',linewidth=1.3,
        marker='o',markersize=8,
        markeredgecolor='white')
```

```
ax.set_xticks(tr,tr_labs)
fig.show()
```

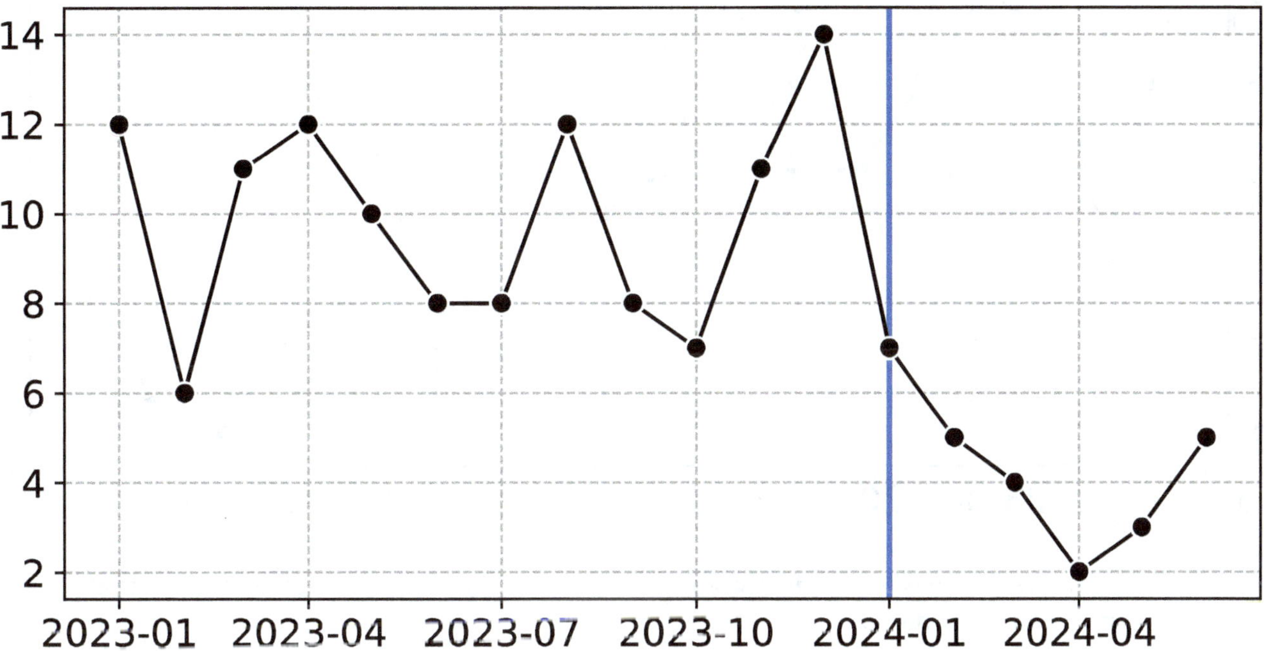

Sometimes the intervention time period can seem ambiguous as to when it occurs. Here it is exactly on January 2024, does that mean the intervention started before January? Sometimes I like to do the plot so the location of the time periods are in the middle of the month. You can use `pd.TimeDelta` to do that.

```
# add 15 days
add15 = pd.Timedelta(15,"D")

fig, ax = plt.subplots(figsize=(8,4))
ax.axvline(x[-6],color='blue',linewidth=2)
# add 15 days
ax.plot(x + add15,y,color='k',linewidth=1.3,
        marker='o',markersize=8,
        markeredgecolor='white')
ax.set_xticks(tr,tr_labs)
fig.show()
```

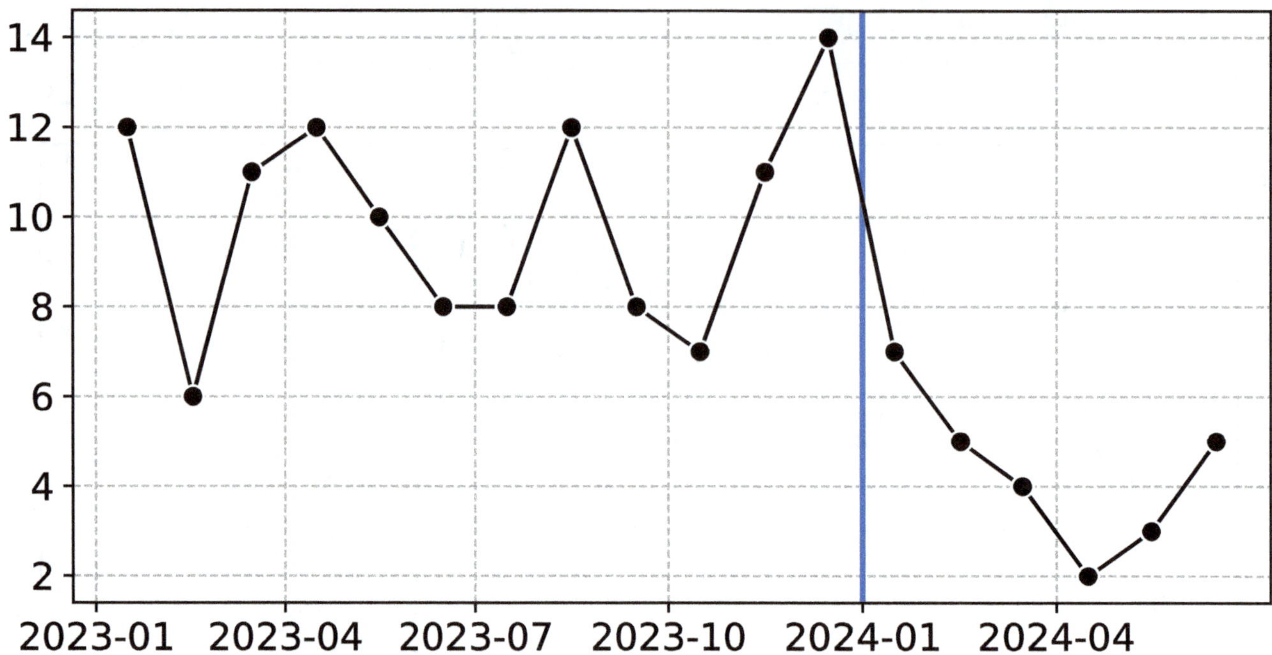

Another approach I like to use is to physically separate the lines. Here there is no ambiguity that January is after the intervention started.

```
fig, ax = plt.subplots(figsize=(8,4))
ax.axvline(x[-6] - add15,color='blue',linewidth=2)
# two seperate lines before/after
ax.plot(x[:-6],y[:-6],color='#888888',linewidth=1.3,
        marker='o',markersize=8,
        markeredgecolor='white',label='Y1')
ax.plot(x[-6:],y[-6:],color='k',linewidth=1.3,
        marker='o',markersize=8,
        markeredgecolor='white',label='Y1')
ax.set_xticks(tr,tr_labs)
fig.show()
```

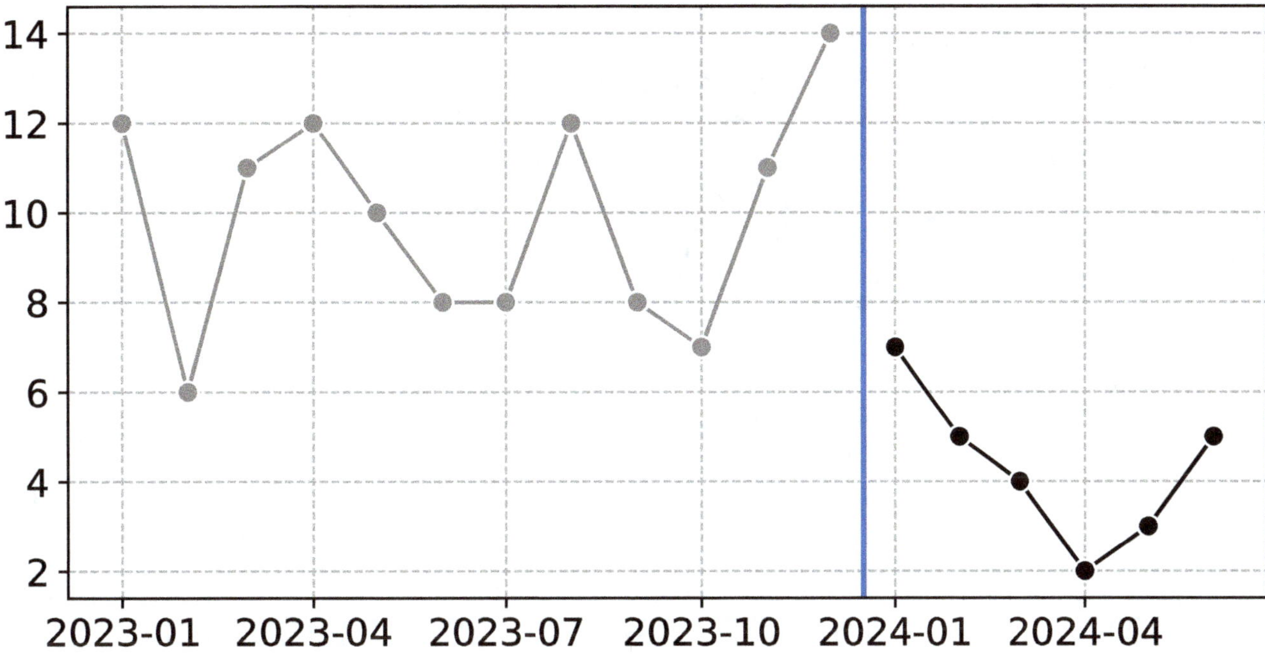

Sometimes you have an entire time span you want to illustrate. To do that, you can use `ax.axvspan`. Note that for both the vertical rule and the area, I plot those *before* the data, so they are by default underneath the actual data values in the plot. Otherwise they would obstruct the data.

```
add5 = pd.Timedelta(5,"D")

fig, ax = plt.subplots(figsize=(8,4))
# polygon
ax.axvspan(x[-12] - add5,x[-6] + add5,color='blue',alpha=0.2)
ax.plot(x,y,color='k',linewidth=1.3,
        marker='o',markersize=8,
        markeredgecolor='white')
ax.set_xticks(tr,tr_labs)
fig.show()
```

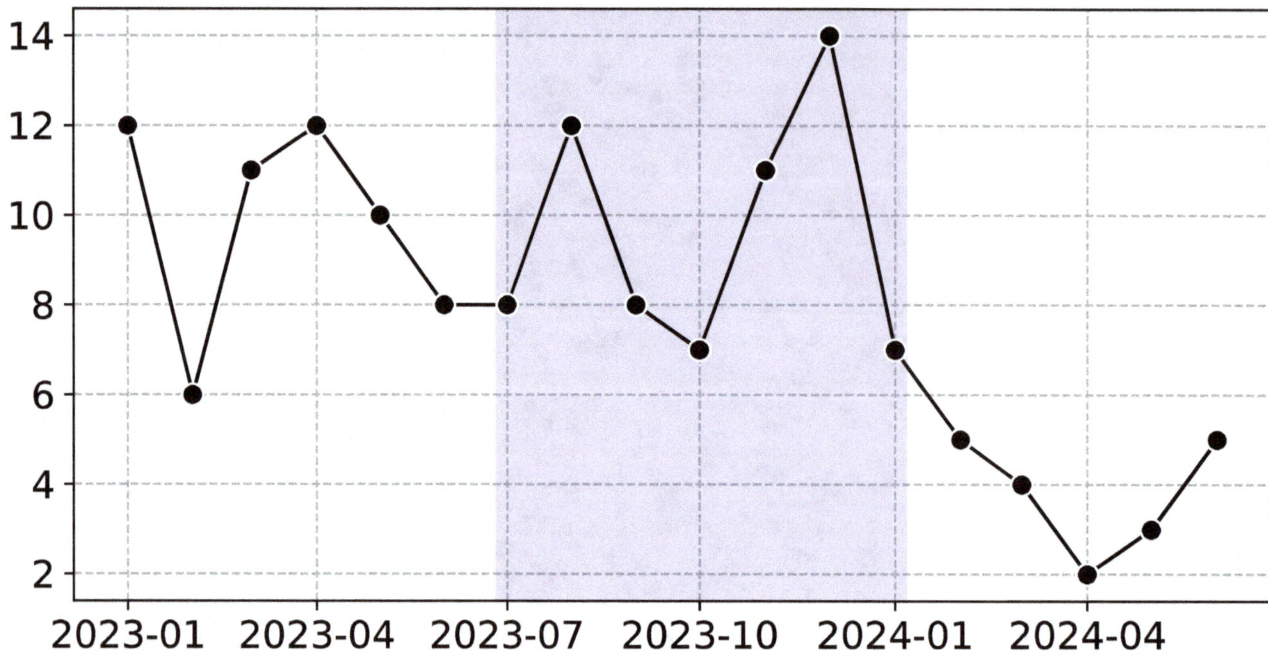

There are similar methods for horizontal rules and areas, `ax.axhline` and `ax.axhspan` respectively.

8.5 Bar plots

Bar plots are another common type of chart. Here is a very simple example

```
label = ['Lab1','Lab2','Lab3']
val = [1,3,2]

fig, ax = plt.subplots()
ax.bar(label,val,color='orange')
fig.show()
```

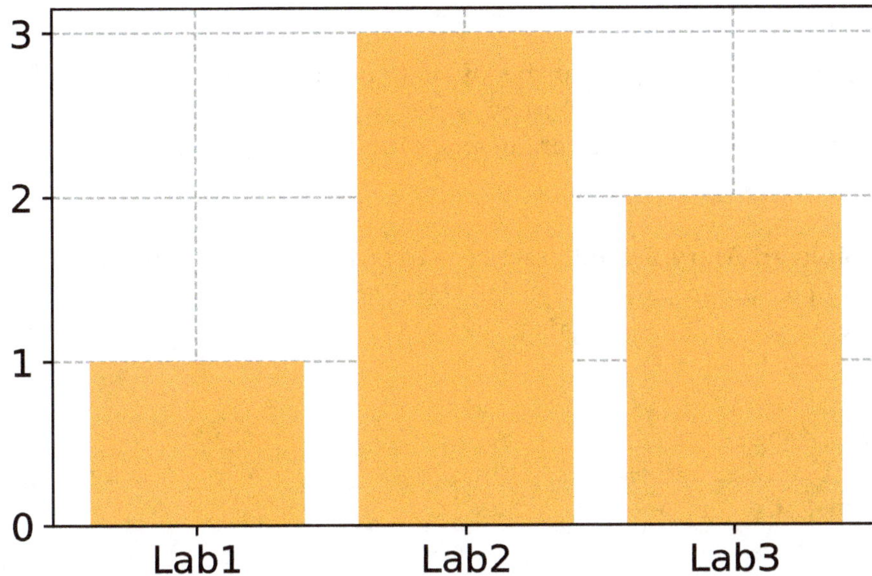

Sometimes if you have longer labels, it is easier to use horizontal bars instead of vertical bars. You can also have line breaks in labels as well. To make horizontal bars, you can use `ax.barh`.

```
label = ['Long label 1','Label across \ntwo lines','Label 3']
val = [1,3,2]

fig, ax = plt.subplots()
ax.barh(label,val,color='orange')
fig.show()
```

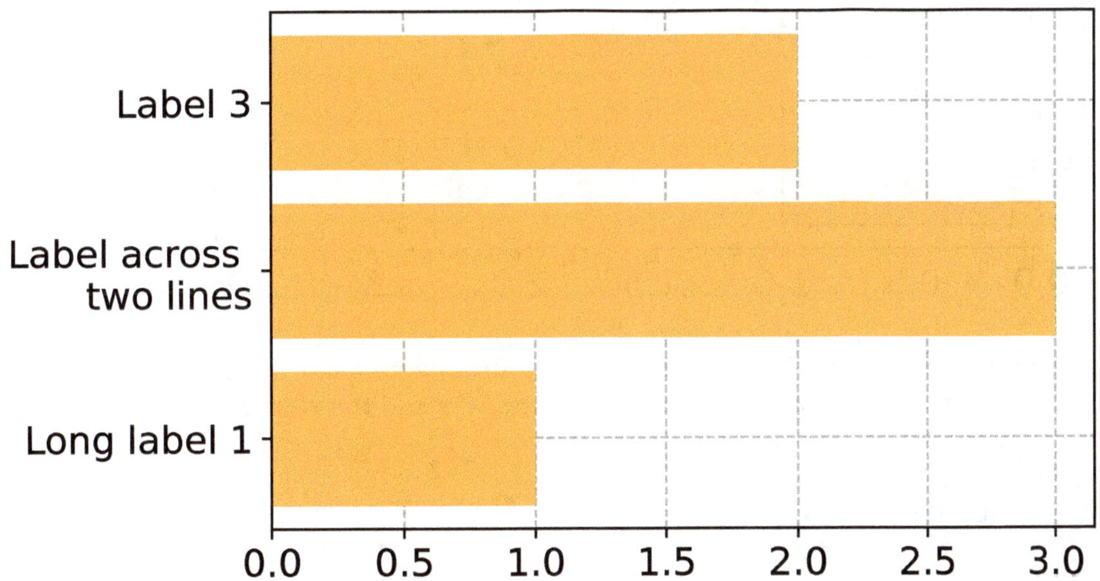

> Note
>
> If running this graph in the REPL, it is quite possible for the labels to be cut-off on the left hand portion of the graph. You can manually resize the chart to be wider to encompass the labels, or click on a button that allows you to reconfigure subplot parameters. One of those parameters is the padding for the labels.

Since in bar charts the label order is often arbitrary, a good practice is to order the data. You can do that via using pandas to order the data, and passing in pandas series to the chart objects. I also like to make outlines for the bars, as this makes them stand out more against the background, you can do this via the `edgecolor` parameter.

```
# turning the data into a pandas df
df = pd.DataFrame(zip(label,val),columns=['lab','val'])
df.sort_values(by='val',inplace=True)

fig, ax = plt.subplots()
ax.barh(df['lab'],df['val'],color='orange',edgecolor='black')
fig.show()
```

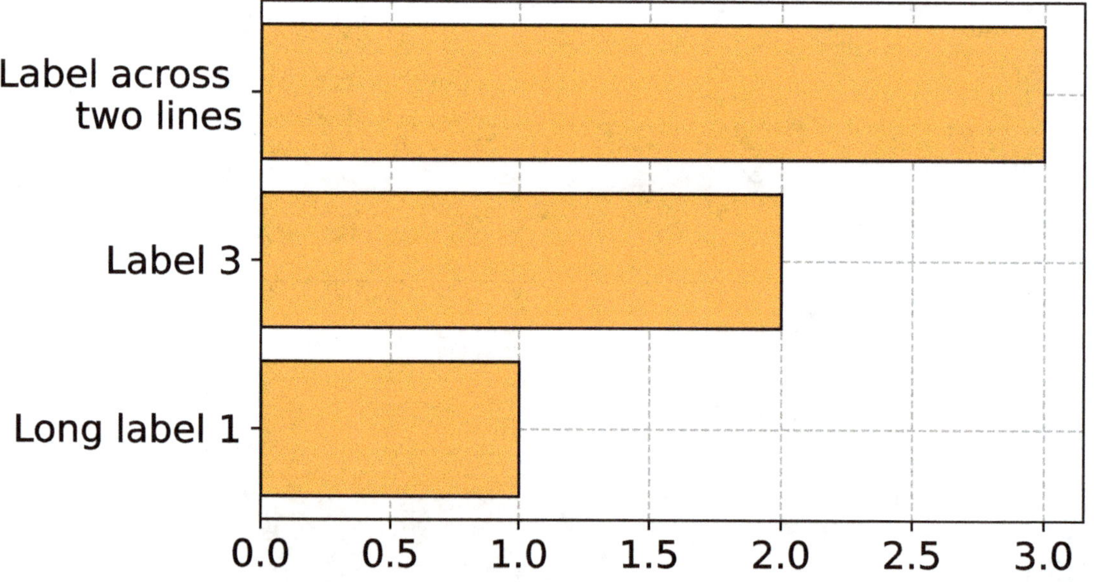

You typically don't need the gridlines for bar charts like this. I also like to make a slight buffer between the edge of the chart, so there is some space at the bottom of the bars and the edge of the plot.

```
# no axis and slight space
fig, ax = plt.subplots()
ax.barh(df['lab'],df['val'],color='orange',edgecolor='black')
ax.grid(False)
ax.set_xlim((-0.05,3.05))
```

```
fig.show()
```

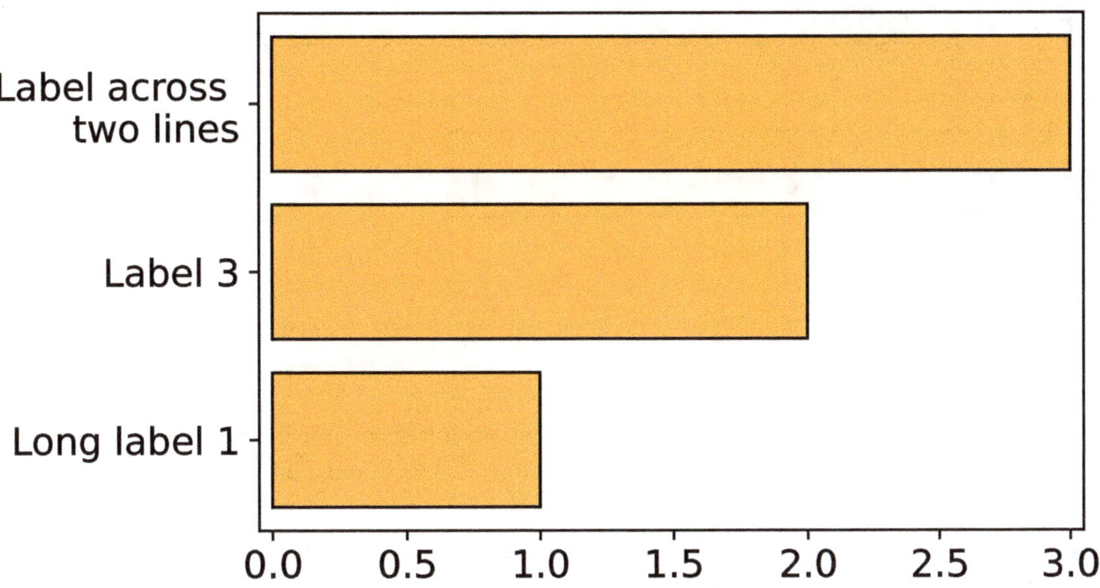

To make clustered or stacked bar charts, you can use plotting methods for pandas dataframes directly. First, I show the same bar plot we created previously using pandas methods directly.

```
# using pandas methods
fig, ax = plt.subplots()

# you pass in the ax as an argument
df.plot.barh(x='lab',y='val',ax=ax,
             color='orange',edgecolor='black')

# the rest is same as before
ax.set_xlabel('') # I don't want "lab" axis label
ax.grid(False)
ax.set_xlim((-0.05,3.05))
fig.show()
```

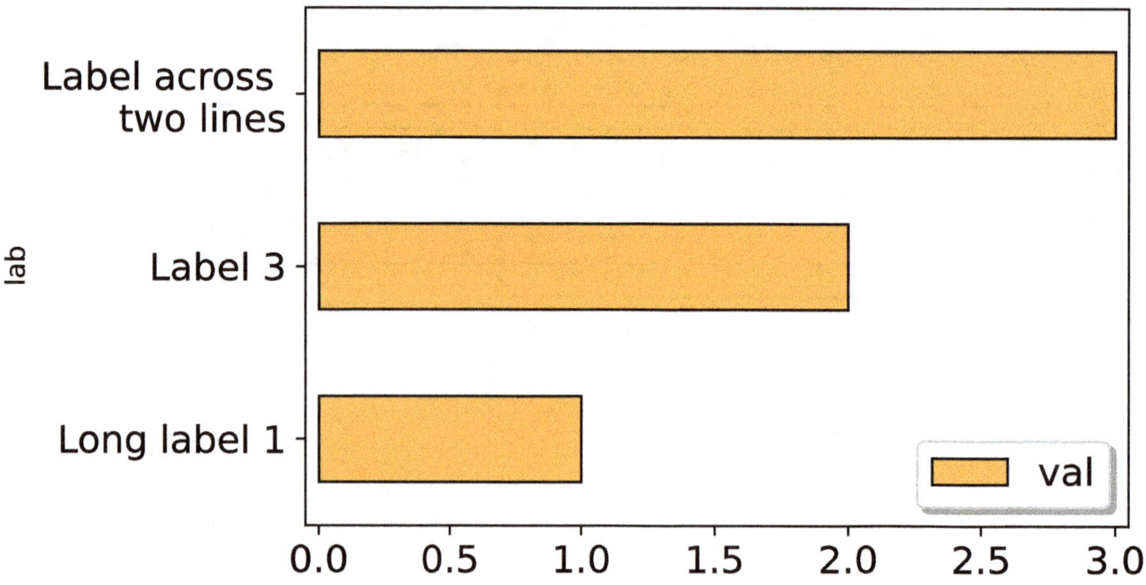

By default this generates a legend. To turn that off you would use `df.plot.barh(...,legend=False)`. But for stacked or clustered bar charts, we will want to include the legend. To make a stacked bar, you can pass in multiple columns and set the index for the dataframe.

```
# making a second column to stack
df['val2'] = [2,1,1]
df.set_index('lab',inplace=True)
df.index.name = None # no axis label

fig, ax = plt.subplots()
# this implicitly plots all the data uses
# index as label and all other columns as bars
df.plot.barh(ax=ax,edgecolor='black',legend='reverse')
ax.grid(False)
ax.set_xlim((-0.05,3.05))
fig.show()
```

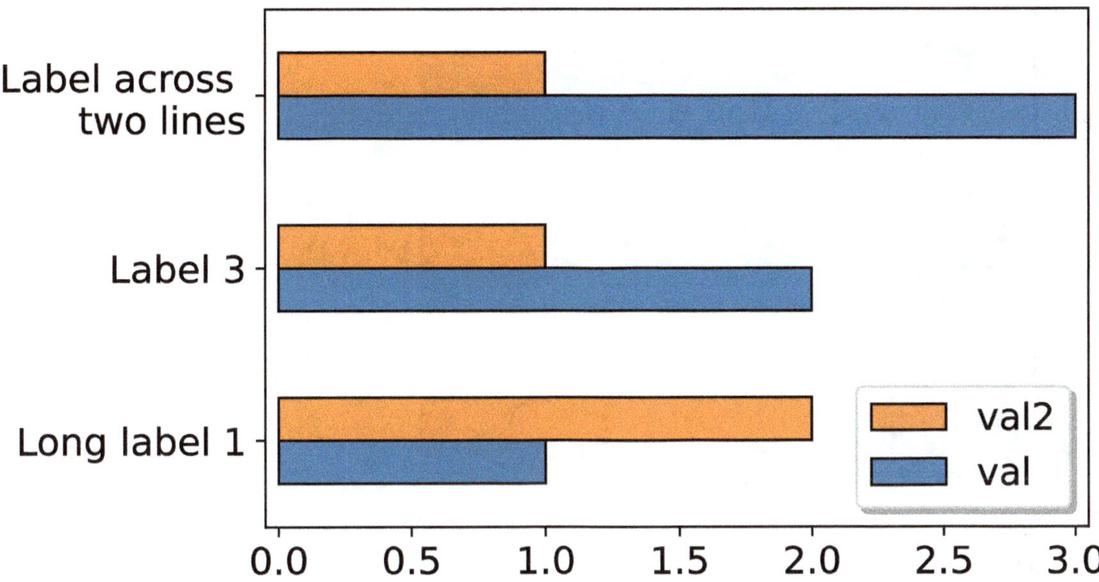

Without `legend='reverse'`, the order of the items in the legend is opposite what you would expect. To make it a stacked bar chart, you can pass in an argument to the pandas method. Here I also get the max for the *sum* across all rows, as that is what is relevant here.

```
# getting max across rows
maxv = df.sum(axis=1).max()

# using pandas methods
fig, ax = plt.subplots()

# this time the legend is in the order I want
df.plot.barh(ax=ax,stacked=True,edgecolor='black')

ax.grid(False)
ax.set_xlim((-0.05,maxv + 0.05))
fig.show()
```

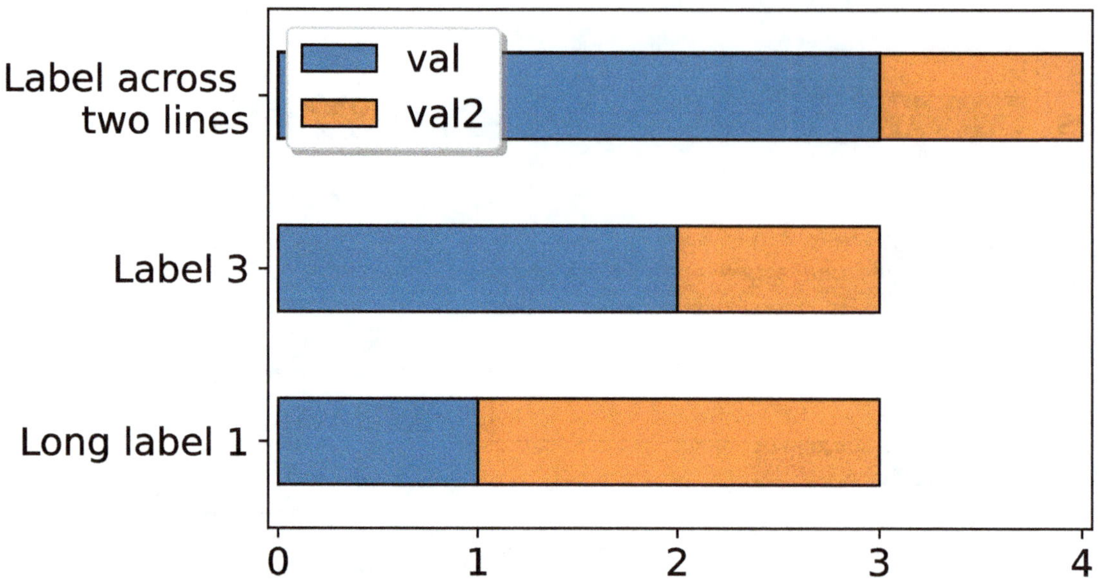

For stacked bars, it is common to normalize the data to sum to 100%. Here we can do that via the pandas dataframe, first by calculating the rowsums, and then using a method to apply division across the entire pandas dataframe. Here I also show creating a percentage formatted set of ticks.

```
# normalizing across rows to sum
# to 100%
rowsum = df.sum(axis=1)
df2 = df.div(rowsum,axis=0)

fig, ax = plt.subplots()
# normalized to sum to 1
df2.plot.barh(ax=ax,stacked=True,edgecolor='black')

# formatting x as percent
x_loc = pd.Series([0,0.25,0.5,0.75,1])
x_str = x_loc.map('{:.0%}'.format)
ax.set_xticks(x_loc,x_str)
ax.grid(False)
ax.set_xlim((-0.05,1.05))
fig.show()
```

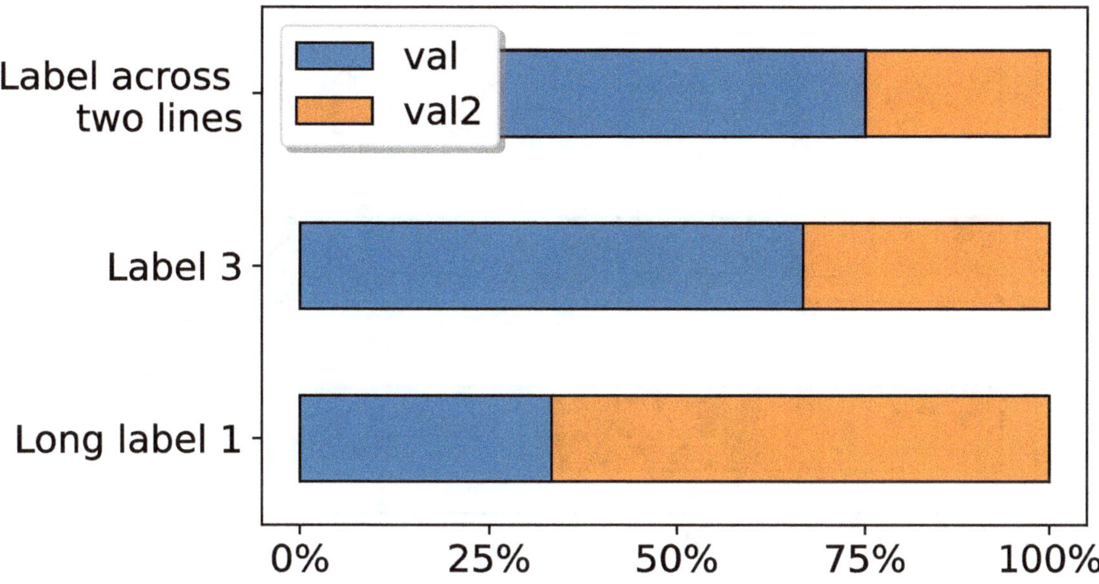

We have pretty much run out of room to place our legend. We can place the legend *outside* of the plot however, by using the `bbox_to_anchor` argument to the `ax.legend` function. Here I set the pandas method to not draw the legend, and then call `ax.legend` in a subsequent command. Also here I show how to pass in specific colors for the bar chart using the pandas method via a dictionary with the column names as keys and colors as values.

```
fig, ax = plt.subplots()

# no legend, but specify colors
cm = {'val': '#ca0020', 'val2': '#0571b0'}
df2.plot.barh(ax=ax,stacked=True,
              color=cm,
              edgecolor='black',
              legend=False)

# places legend to the right of plot
ax.legend(bbox_to_anchor=(1.0, 0.5))
ax.grid(False)
ax.set_xlim((-0.05,1.05))
fig.show()
```

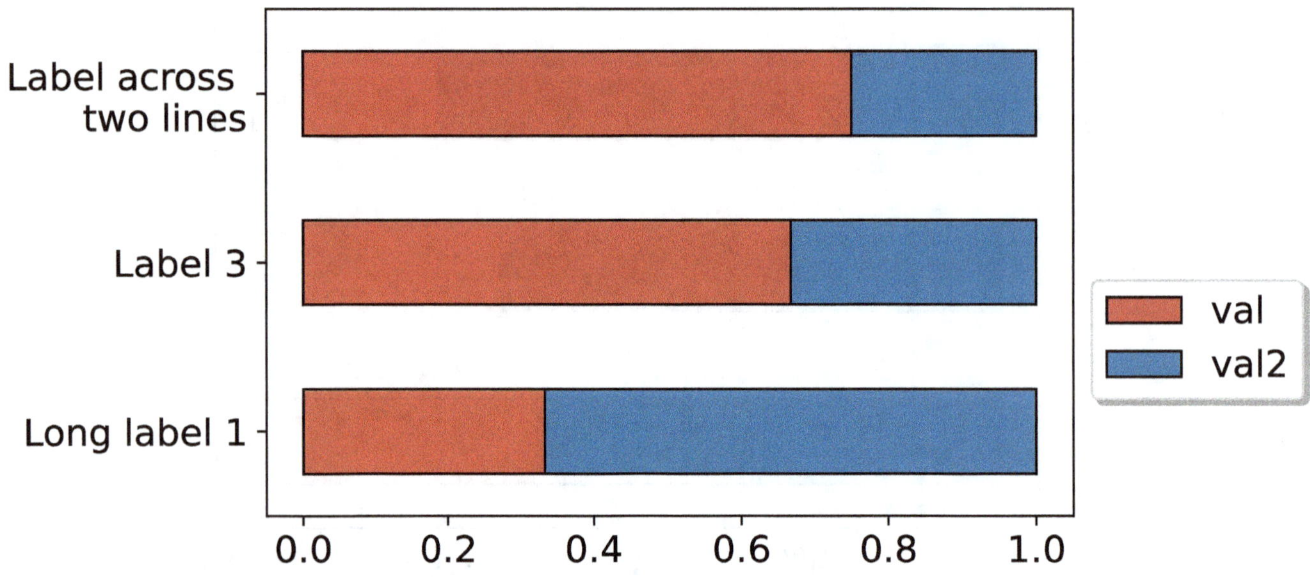

Sometimes placing the legend above or below the plot is preferable instead of off to the side. Here is an example plotting the legend above the plot area, and making the legend multiple columns.

```
fig, ax = plt.subplots()

# no legend
df2.plot.barh(ax=ax,stacked=True,
              color=cm,
              edgecolor='white',
              legend=False)

# places legend to the right of plot
ax.legend(bbox_to_anchor=(0.3, 1.01),ncol=2)
ax.grid(False)
ax.set_xlim((-0.05,1.05))
fig.show()
```

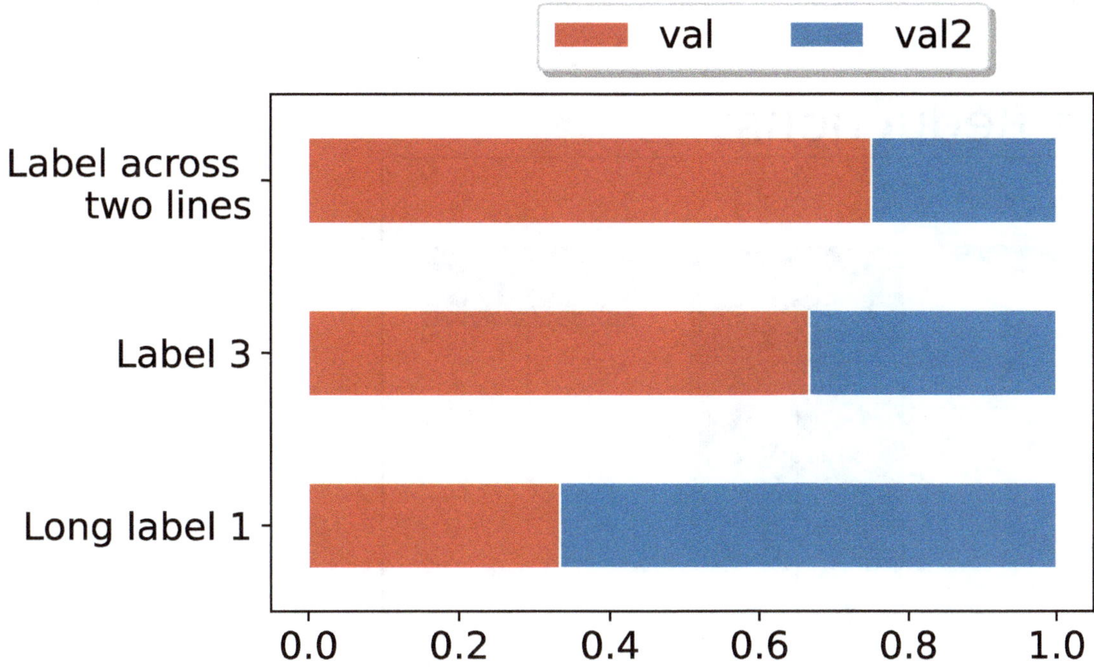

If you wanted it below the plot, you would do something like

```
ax.legend(bbox_to_anchor=(0.3,-0.1),...)
```

The arguments to the bbox parameter are the `(x,y)` coordinates for where to place the legend relative to the graph area. So a negative value for the y coordinate means below the plot, and a positive value means above the plot. Ditto for placing the legend to the left or the right of the graph.

One final example I want to show are superimposing error bars on a plot. Say you did an experiment, and showed that in three areas you had different effects of reducing crime over time, so here we have *negative* bars.

```
# Point estimates and errors
lab = ['Area1','Area2','Area3']
pr = np.array([-1,-20,-5])
error_low = np.array([4,5,5])
error_high = np.array([5,10,5])

# this is how error bars need to be
# passed in matplotlib
err = np.array([error_low,error_high])

# barplot with error bars
fig, ax = plt.subplots()
ax.bar(lab,pr,yerr=err)
ax.set_title("Crime Reductions")
```

```
fig.show()
```

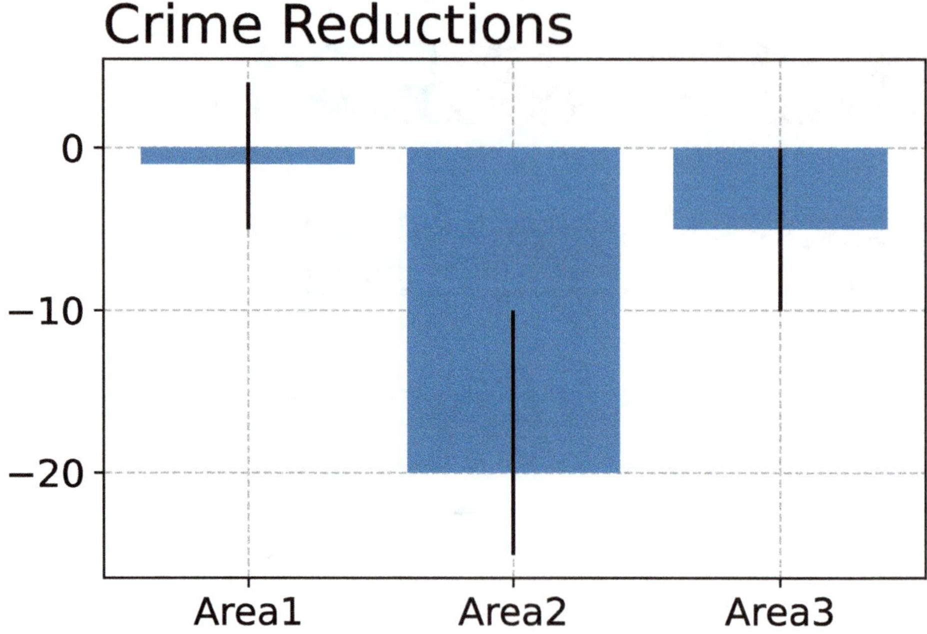

> Note
>
> I have created simple metrics to determine if a place based experiment is reducing crime, the weighted displacement difference (WDD) test, just by comparing before and after crimes counts in treated and control areas. Check out https://crimede-coder.com/graphs/WDD for a free online tool and documentation on that test.

I don't tend to like those types of charts though. Often times it is easier to just plot the point estimate directly. Here you can use `ax.errorbar` to do that.

```
# points with error bars
fig, ax = plt.subplots()
ax.errorbar(lab,pr,yerr=err,
            fmt='s',
            ecolor='k',
            markerfacecolor='k',
            markeredgecolor='white',
            markersize=14)
ax.set_title("Crime Reductions")
fig.show()
```

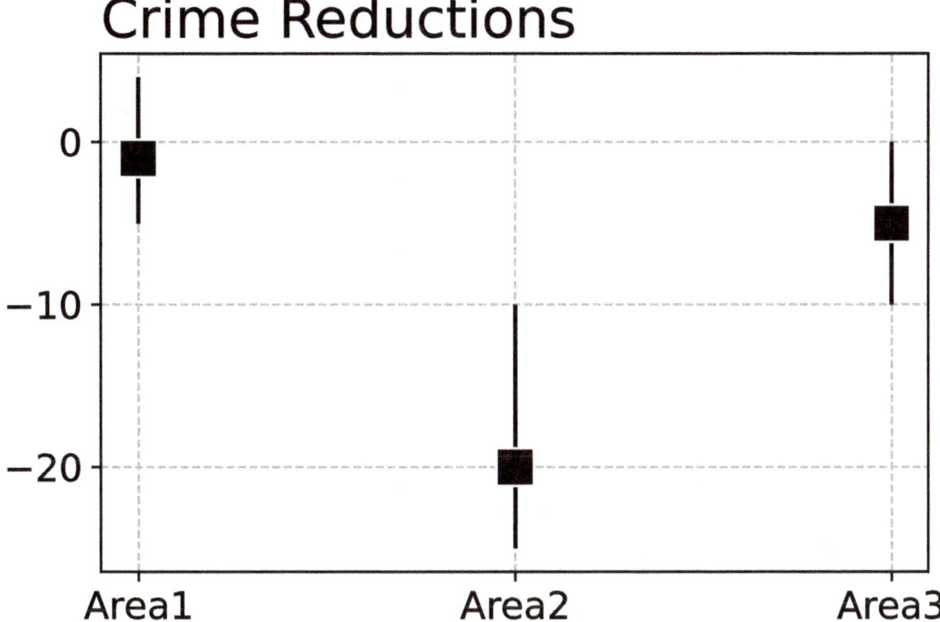

A common critique of bar plots is that you should always have the origin of the plot be at zero, as you are intended to interpret *the length* of the bar in the plot. On occasion people will use bar charts that are not centered on zero to try to trick viewers into thinking differences are bigger than they really are (such as political polling for different candidates, and the bars start at 45% instead of 0). Using points and error bars though avoids that problem all together, so people only focus on the final point estimates and their errors, not the length of bars.

8.6 Scatterplots

Scatterplots display points, you can use `ax.plot` and pass a third (unnamed) argument after the x and y vector arrays to simply display a particular point marker without the line.

```
x = [1,2,3]
y = [0.5,1.2,0.8]
y2 = [1.1,0.9,0.7]

# can use a third ordered arg to make
# just a scatterplot
fig, ax = plt.subplots()
ax.plot(x,y,'o',markersize=12)
ax.plot(x,y2,'s',markersize=12)
fig.show()
```

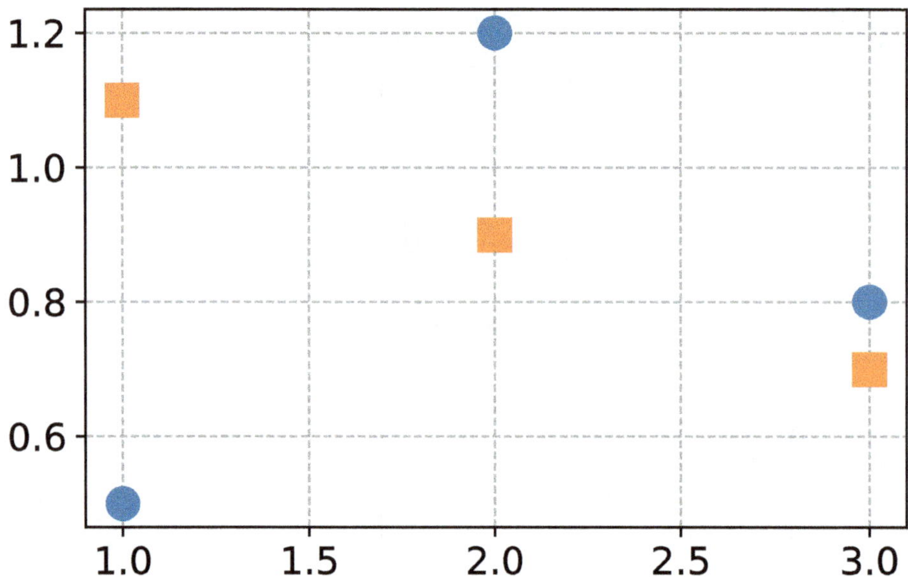

If you use this format, you *need* to pass in the arguments in a particular way. You can however use a different axes method, `ax.scatter`, to also produce points. It has many of the same types of arguments to manipulate the marker size, color, etc. But unfortunately many of the keyword arguments for `ax.scatter` are different than `ax.plot`.

```
# can also use ax.scatter
fig, ax = plt.subplots()
ax.scatter(x,y,s=12**2,marker='o',edgecolors='k')
ax.scatter(x,y2,marker='*',s=14**2,c='red')
fig.show()
```

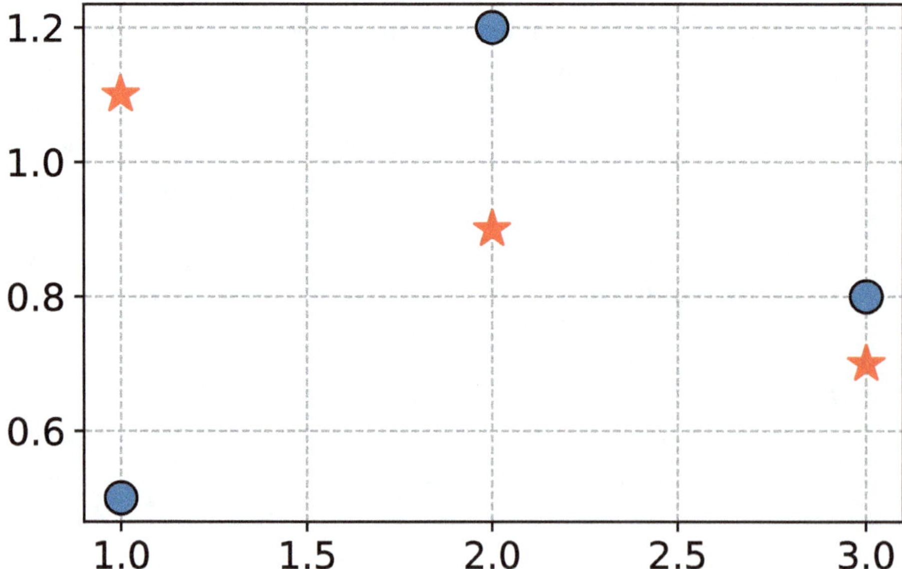

For the sizes, I pass in the arguments as `12**2`, because by default they try to map to areas when scaling the size of the markers when using `ax.scatter`. Here I also show using a triangle symbol, I often only use symbols `o` (for circle), `s` (for square), and `*` (for star). If you do an internet search for `matplotlib markers` you can find a list of up to date symbols.

> Note
>
> I often use google to look up help, instead of doing `help(ax.scatter)`. Matplotlib has nice documentation online, see https://matplotlib.org/stable/api/markers_api.html#module-matplotlib.markers for example for markers. But it is likely that urls change, so I try to avoid giving links in the book to documentation.

When using `ax.scatter`, you can also have the points be varying sizes and colors.

```
# sequential values
x = range(10)
y = np.linspace(1,30,10)

fig, ax = plt.subplots()
ax.scatter(x,y,s=y**2,
           marker='o',
           c=y,
           edgecolor='k')
fig.show()
```

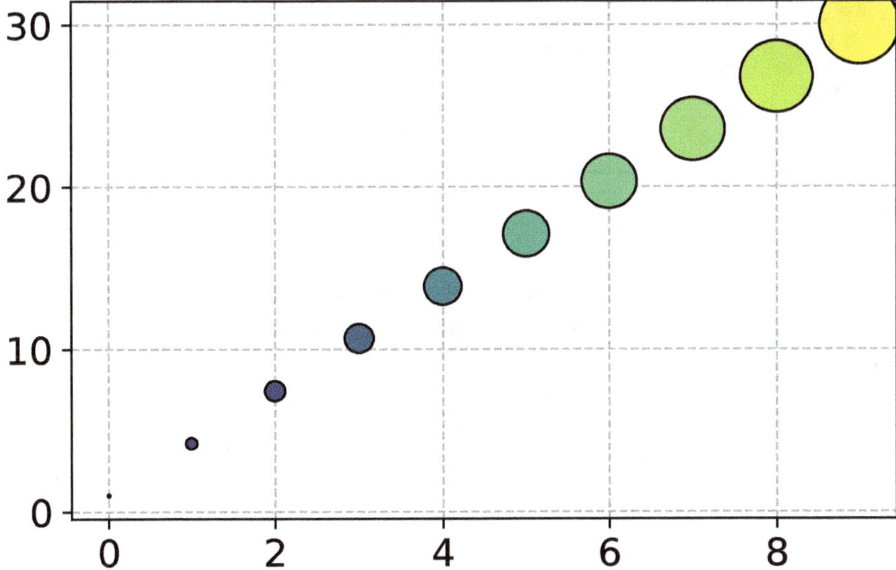

Here I map the y axis value to both sizes and colors, but they can be different vectors. Here I also show off a different colormap, colorbrewer's green palette.

```
# random data that overlaps
n = 100
x = np.random.rand(n)
y = np.random.rand(n)
cl = np.random.rand(n)
si = np.random.rand(n)


fig, ax = plt.subplots()
ax.scatter(x,y,s=(si*30)**2,
           marker='s',
           c=cl,
           cmap='Greens')
fig.show()
```

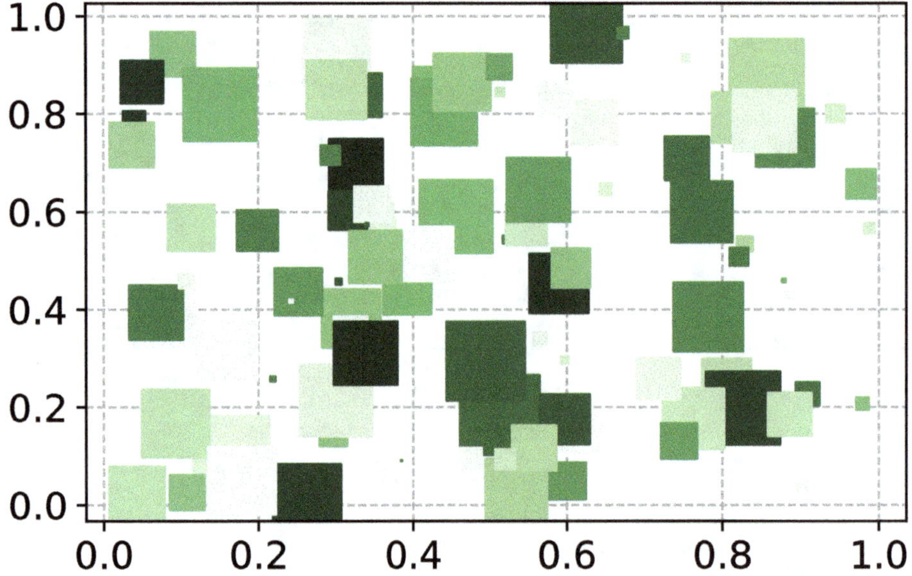

Previously I have shown creating a *discrete* color scale for different elements in a plot. This is different, in that it is a continuous color scale for numeric values. So there is a different method to create a legend in this scenario, a *colorbar* in matplotlib lingo.

```
fig, ax = plt.subplots()
cbar = ax.scatter(x,y,s=(si*30)**2,
           marker='s',
           c=cl,
           cmap='plasma',
           edgecolor='k')
fig.colorbar(cbar)
fig.show()
```

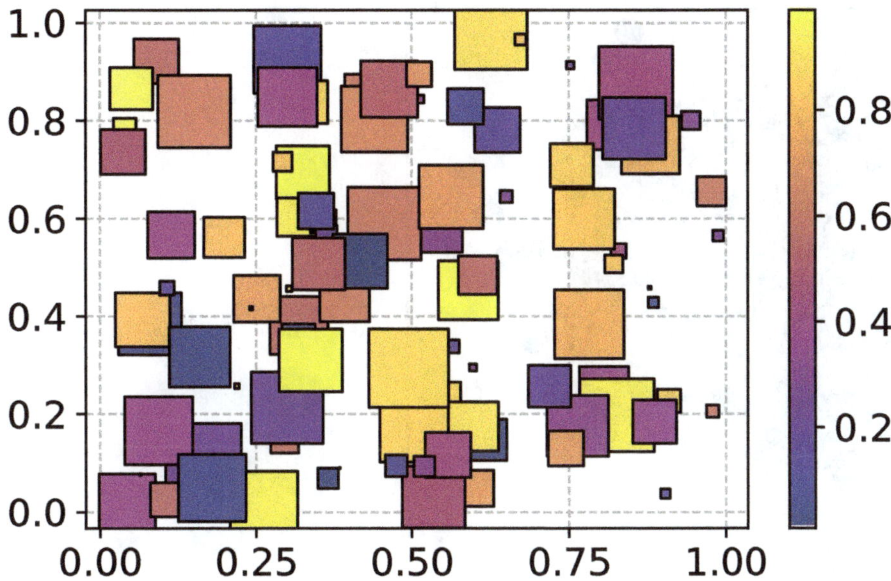

For scatterplots in which elements can overlap, I like drawing the point elements with an edgecolor, as that can help disambiguate elements that overlap. Without a border, they tend to blend together into each other. But for very dense scatterplots, even with drawing the borders, the overplotting becomes an issue in interpreting any patterns.

```
# random normal data
n = 5000
x = np.random.normal(size=n)
y = x*0.2 + np.random.normal(size=n)

fig, ax = plt.subplots()
ax.scatter(x,y,s=10**2,
           marker='o',
           edgecolor='k')
fig.show()
```

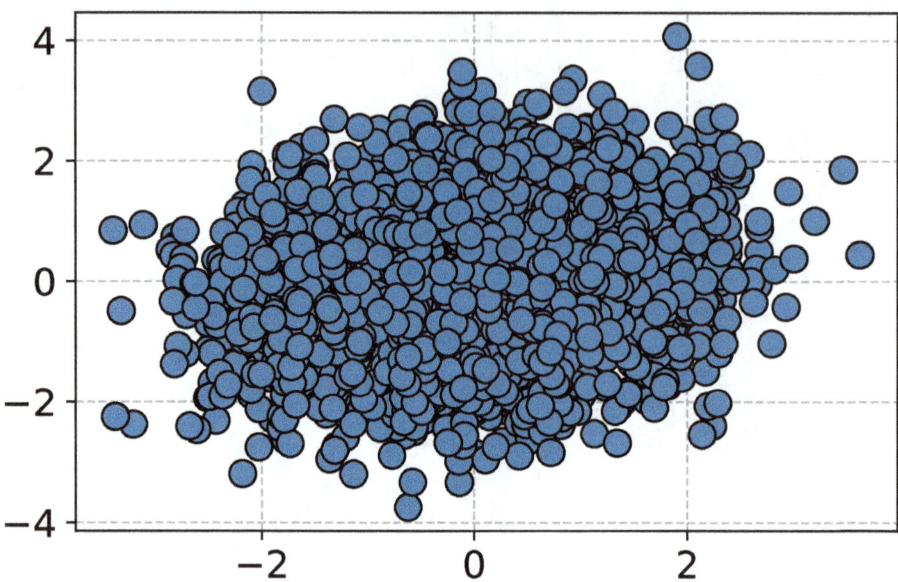

One way to solve this is to make the points be tinier and transparent. The marker `.` is a quick way to do this.

```
# making tinier and transparent
fig, ax = plt.subplots()
ax.scatter(x,y,marker='.',
           alpha=0.2)
fig.show()
```

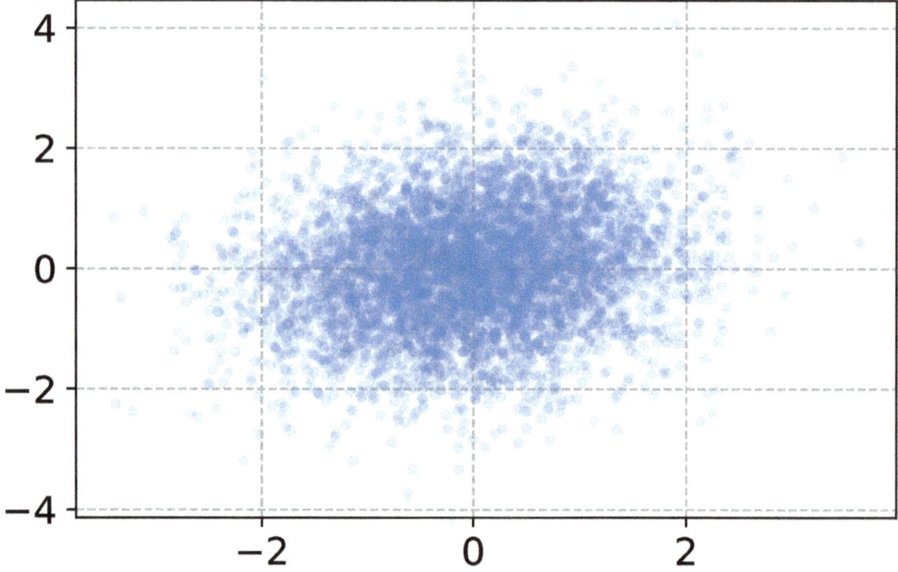

That helps some, but it is still difficult to really see patterns in this scatterplot. Another approach is to *bin* data, and display a polygon showing the number of observations within that area. Matplotlib has a convenience function for aggregating counts to hexagons.

```
# hexbinning
fig, ax = plt.subplots()
cbar = ax.hexbin(x,y,
          gridsize=20,
          edgecolors='grey',
          cmap='PuBu',
          mincnt=1)
fig.colorbar(cbar)
fig.show()
```

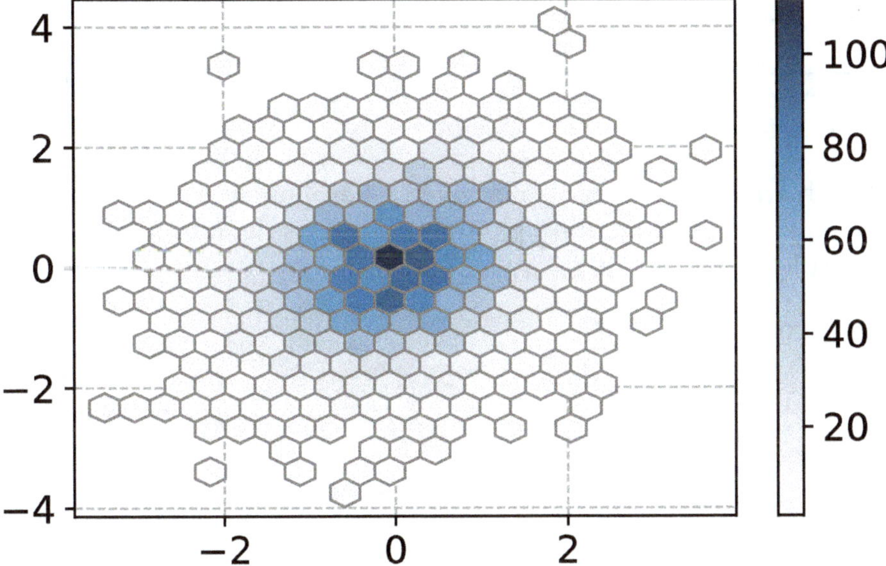

You can manipulate the gridsize to make it smaller, and also make the borders smaller as well (if you omit the edgecolor parameter, it draws the hexagons with no edges at all).

```
# finer grid
fig, ax = plt.subplots()
cbar = ax.hexbin(x,y,
          gridsize=40,
          linewidth=0.1,
          edgecolors='white',
          cmap='Reds',
          mincnt=1)
fig.colorbar(cbar)
fig.show()
```

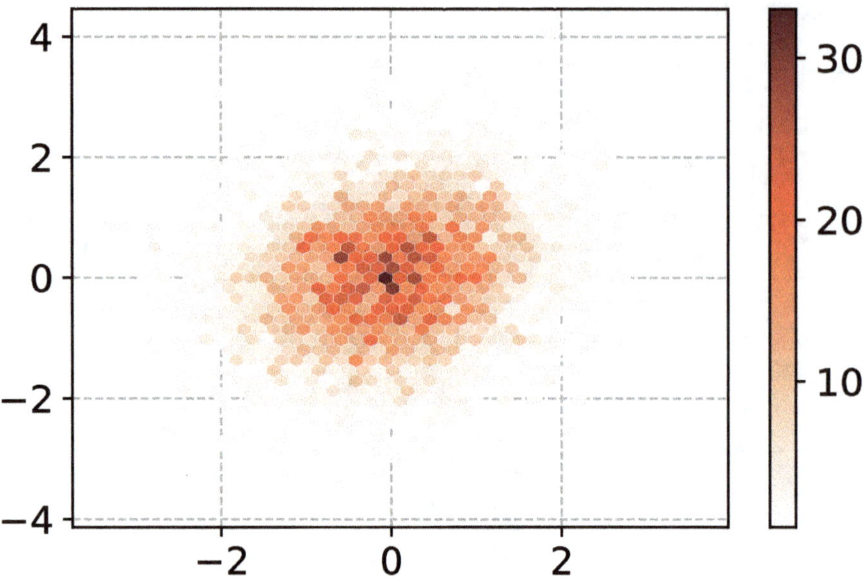

Here I have been showing the plot with areas with 0 points not given a hexagon at all. You may wish to cover the entire plot though, just omit the prior `mincnt` parameter, and its default is to include a tesselation of hexagons over the entire plot.

```
# finer grid
fig, ax = plt.subplots()
cbar = ax.hexbin(x,y,
          gridsize=50,
          cmap='inferno')
fig.colorbar(cbar)
fig.show()
```

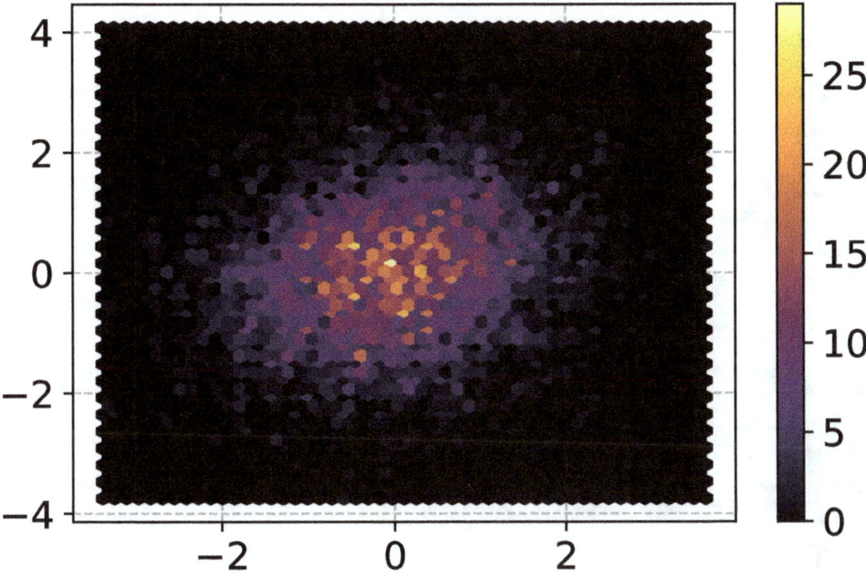

> Note
>
> There are other ways to do similar type plots. You can bin using squares instead of hexagons, or you can do a bivariate kernel density estimate. The *seaborn* library has a convenient set of functions to create bivariate KDE plots. I believe the binned counts though are often easier to interpret.

8.7 Examining distributions

The final examples I want to show in this chapter are examining entire distributions, as well as making several different plots at once. Using pandas methods is the most convenient way in my opinion to create a histogram.

```
# creating data that
# is a mixture
x1 = np.random.normal(3.5,1,1000)
x2 = np.random.normal(6,1,500)
group = ['A']*1000 + ['B']*500
df = pd.DataFrame(zip(np.concatenate([x1,x2]),group),
                  columns=['Val','Gr'])

# histogram with pandas
fig, ax = plt.subplots()
df['Val'].hist(ax=ax)
fig.show()
```

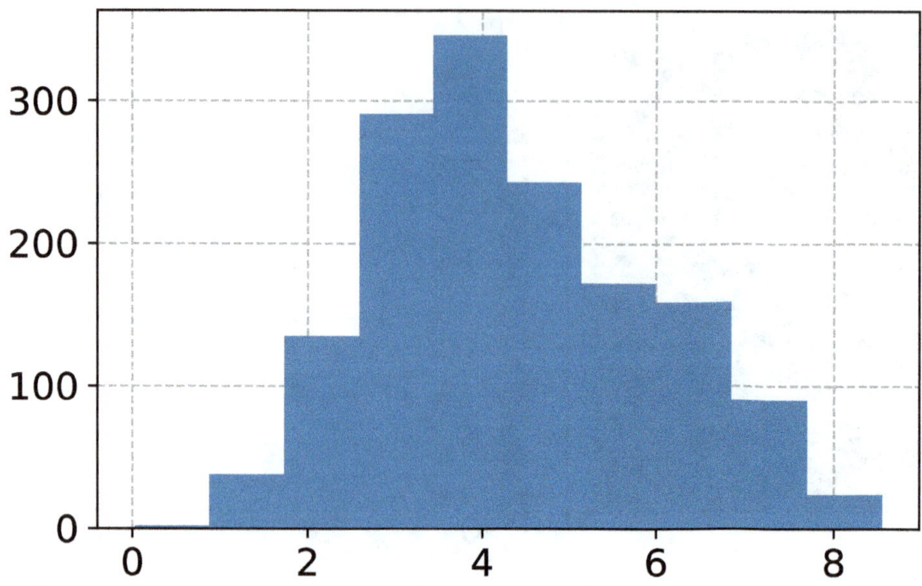

The default number of bins for many situations is too course. So I often up the number of bins (50 to 100 bins works well in my experience). This better illustrates the mixture data I simulated with two different distributions in this example, resulting in a bi-modal distribution.

```
# more bins
fig, ax = plt.subplots()
df['Val'].hist(ax=ax,bins=50,alpha=0.8)
fig.show()
```

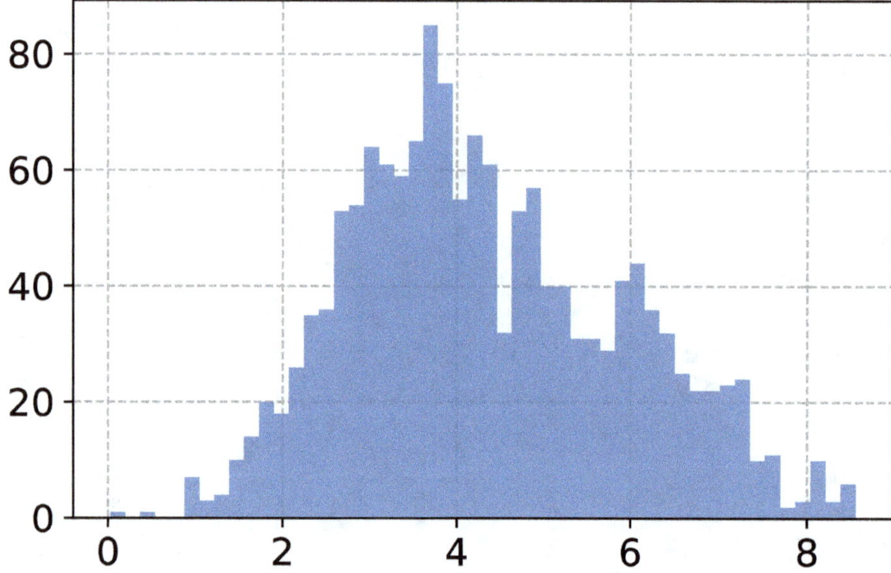

Note

In the PDF version of the book, you can sometimes see border edges for the histogram. This is partially an issue with how PDF renders polygons with transparency, and does not occur when saving the image to a png format. By default, histograms typically don't show the borders of the bars. If you want to show the borders, you can use `.hist(...,ec='k')` to show the borders of the bars as black for example.

A trick I have not shown is that in some pandas plotting methods, you can use `groupby` to split the data and draw multiple elements. Here is an example of that with histograms.

```
# plotting each group
fig, ax = plt.subplots()
df.groupby('Gr')['Val'].hist(bins=40,
    ax=ax,legend=True)
fig.show()
```

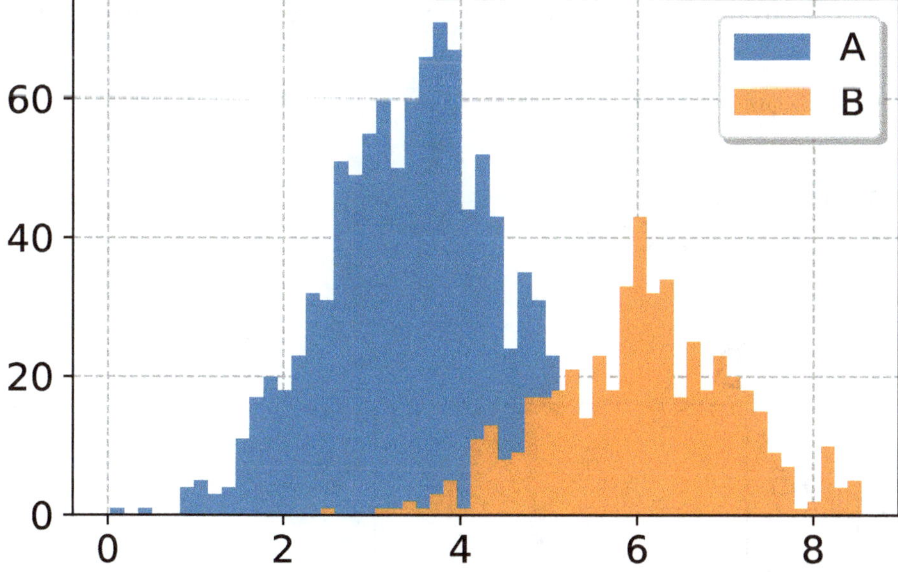

You can see that group B has a smaller overall volume. If you want the two histograms to be more apples to apples in overall distribution, you can display the *density*, instead of counts on the y-axis. This makes it so the area covered by each subgroup is the same (and should sum to 1). Also displaying as semi-transparent using `alpha` allows you to see the grid lines behind the areas, as well as the overlapping areas.

```
# plotting density instead of counts
fig, ax = plt.subplots()
df.groupby('Gr')['Val'].hist(bins=40,alpha=0.8,
    ax=ax,legend=True,density=True)
fig.show()
```

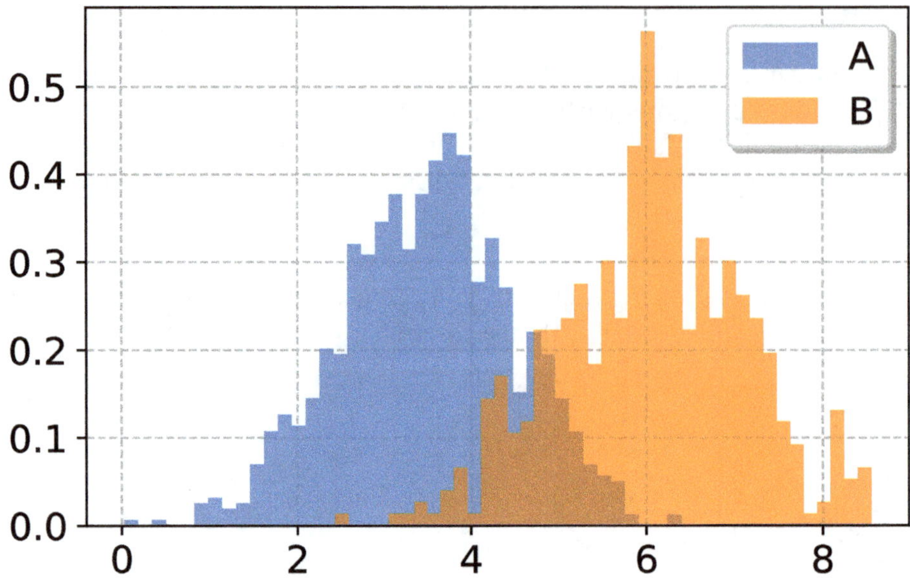

Instead of superimposing the histograms all in the same figure area, you can create *small multiple* plots via a `by` argument.

```
# small multiple
fig, ax = plt.subplots(1,2)
df['Val'].hist(bins=40,alpha=0.8,
         ax=ax,by=df['Gr'],density=True)
fig.show()
```

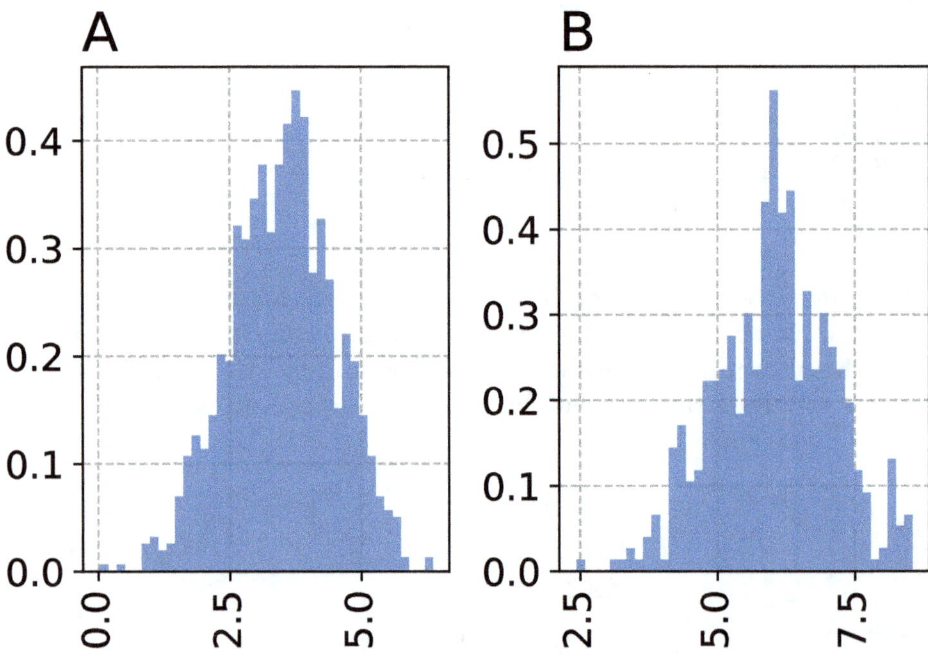

This shows off something new I have not shared before, `plt.subplots(1,2)`. What this does is creates multiple subplots, one row and two columns of graphs. These by default do not share axes, but you can change that to have them have shared axes in either the x or y dimension. Here I also show generating the charts so it is two rows and one column of graphs.

```
# small multiple stacked
fig, ax = plt.subplots(2,1,sharex=True,sharey=True)
df['Val'].hist(bins=40,alpha=0.8,ax=ax,
               by=df['Gr'],density=True)
fig.show()
```

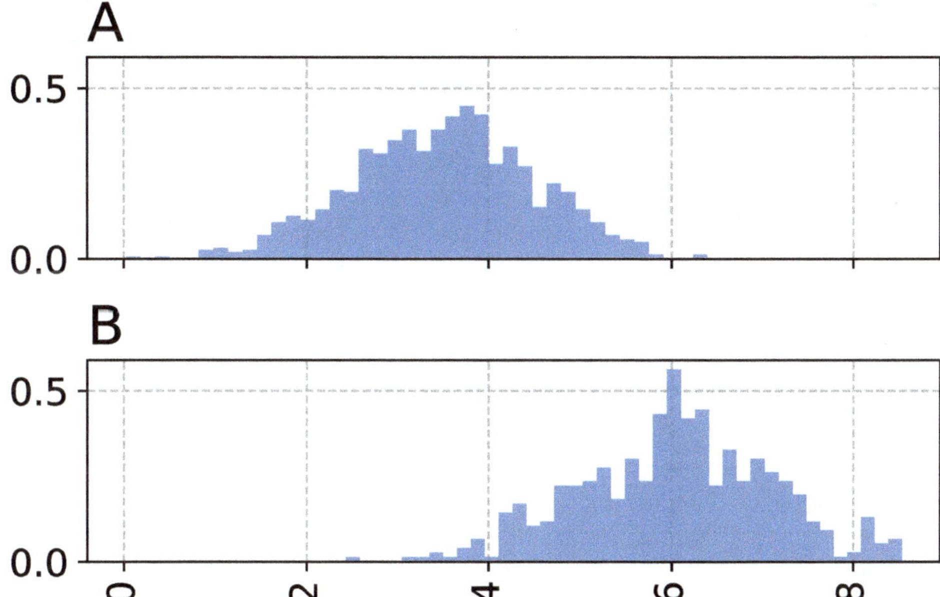

This uses the pandas methods to fill in each of the subplots, but you can fill in each subgraph like it is a totally independent graph. Here I create 4 subgraphs in a two by two grid. I however only fill in three of the subgraphs, and leave the graph on the lower right blank.

```
# three different charts, ax here is an array
fig, ax = plt.subplots(2,2,figsize=(8,8))

# superimposed histogram top left
df.groupby('Gr')['Val'].hist(bins=40,alpha=0.8,
   ax=ax[0,0],legend=True,density=True)

# Boxplots on top right, need data in a particular
# wide way to do this
grv = []
dat = []
for gr, data in df.groupby('Gr')['Val']:
```

```
    grv.append(gr)
    dat.append(data)

ax[0,1].boxplot(dat,vert=True,labels=grv)

# CDF plot on bottom left
df.groupby('Gr')['Val'].hist(bins=40,alpha=0.8,
   ax=ax[1,0],legend=False,density=True,
   histtype="step",cumulative=True)

# Clear the chart on the bottom right
ax[1,1].axis('off')

fig.show()
```

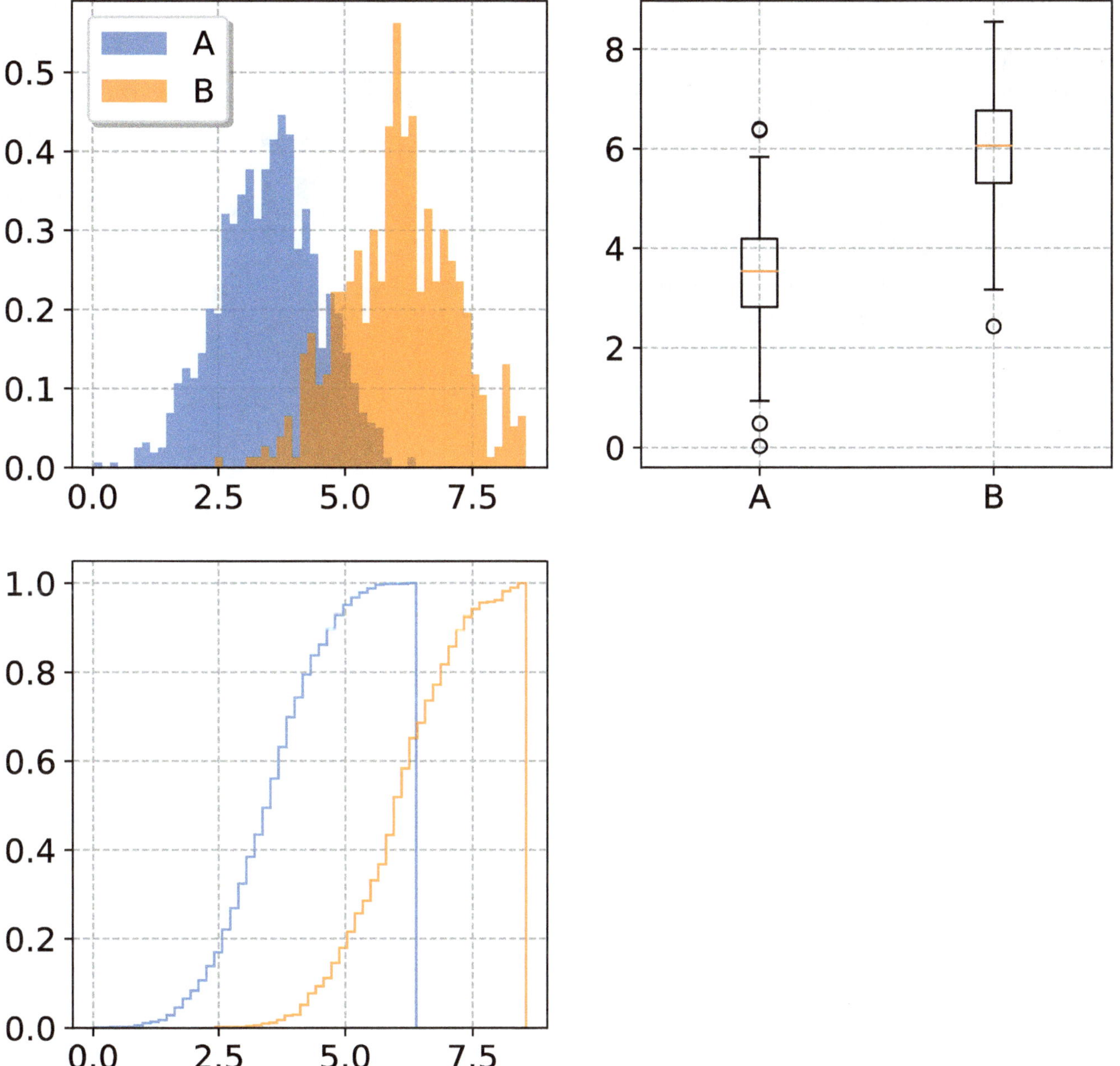

There is a way to make each subplot though in this framework take up a different amount of space vertically or horizontally using `subplot_mosaic`. Here I stretch out the bottom graph across the entire figure area.

```
# bottom chart spans the whole x chart range
mos = [['A','B'],
       ['C','C']]

fig, ax = plt.subplot_mosaic(mos,figsize=(8,8))
```

```
# note access each sub-axis via letter in mos
df.groupby('Gr')['Val'].hist(bins=40,alpha=0.8,
   ax=ax['A'],legend=True,density=True)

ax['B'].boxplot(dat,vert=True,labels=grv)

# CDF plot stretched whole way
df.groupby('Gr')['Val'].hist(bins=40,alpha=0.8,
   ax=ax['C'],legend=False,density=True,
   histtype="step",cumulative=True)

fig.show()
```

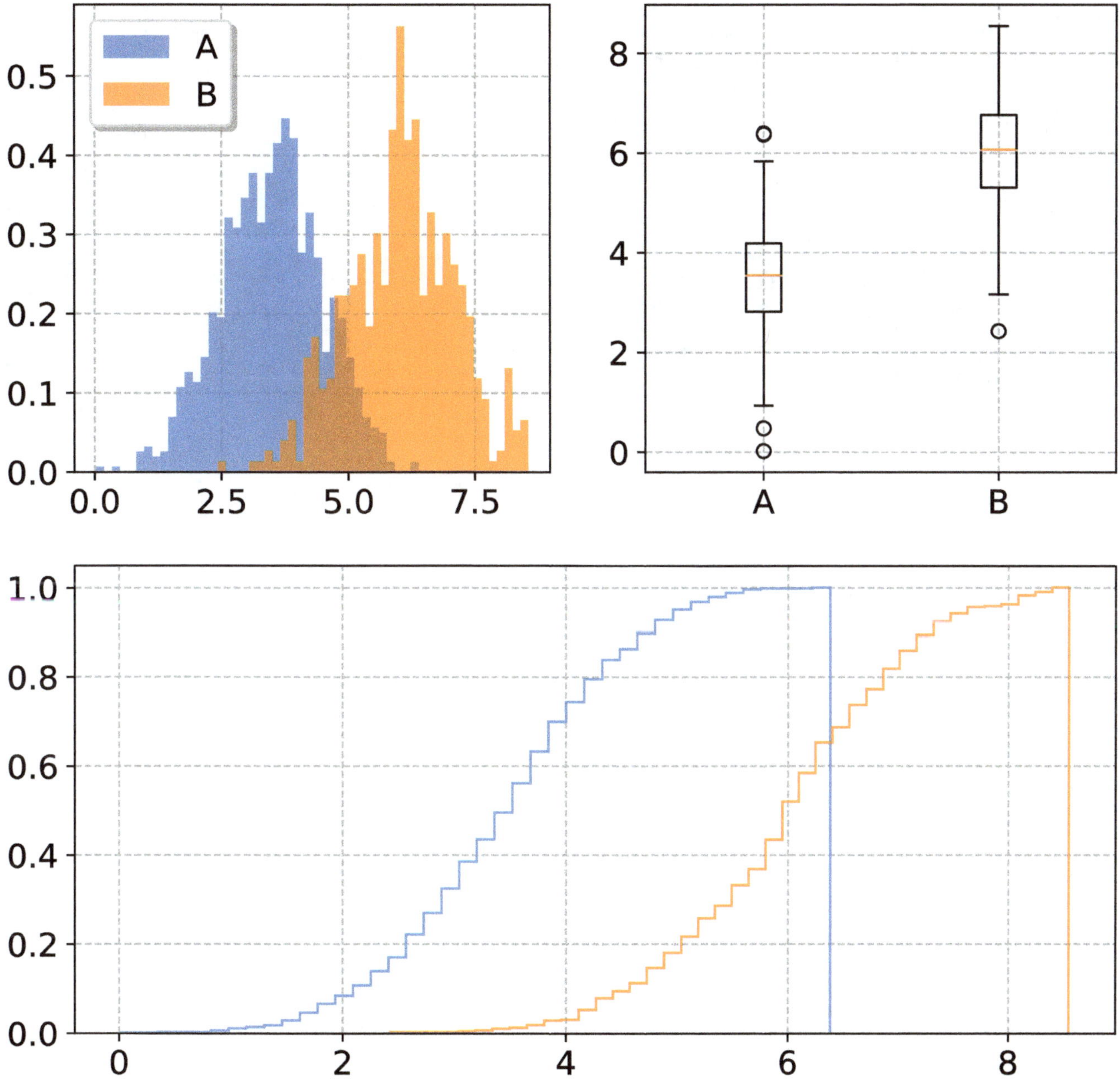

You could also have charts take up more vertical space, here I show the boxplot taking up two rows of space, but less horizontal space.

```
mos = [['A','B'],
       ['C','B']]

fig, ax = plt.subplot_mosaic(mos,
                             width_ratios=[2, 1],
                             figsize=(8,8))
```

```
df.groupby('Gr')['Val'].hist(bins=40,alpha=0.8,
   ax=ax['A'],legend=True,density=True)

# now this takes up two rows
ax['B'].boxplot(dat,vert=True,labels=grv)

# CDF is not stretched
df.groupby('Gr')['Val'].hist(bins=40,alpha=0.8,
   ax=ax['C'],legend=False,density=True,
   histtype="step",cumulative=True)

fig.show()
```

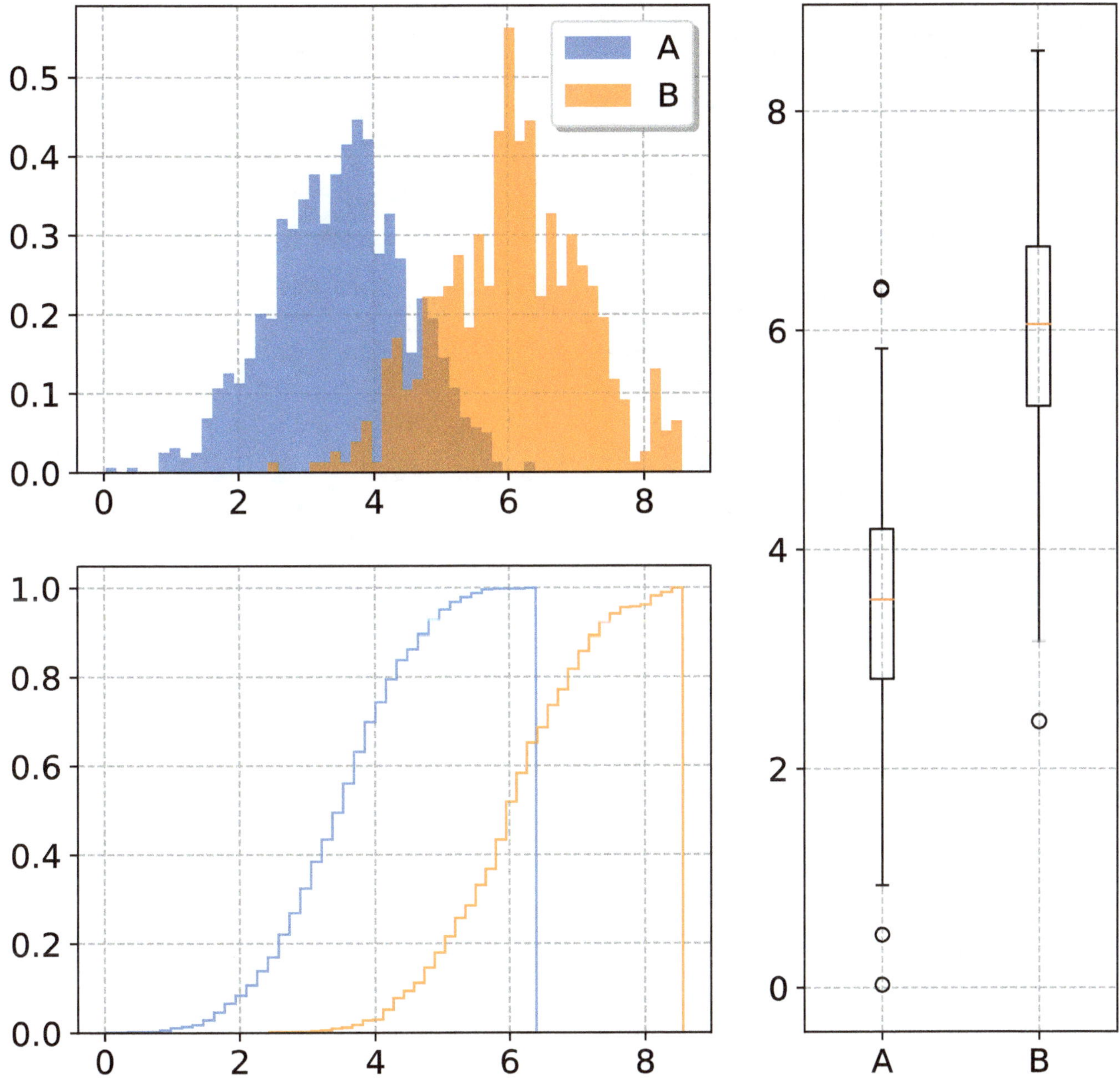

There are ultimately *many* different types of plots I have not illustrated in this chapter. Maps, conditional regression plots, sankey diagrams, treemaps, timelines, network graphs, etc. This chapter does however cover much of the regular functionality I have a need for in creating regular graphs. Many of those additional diagrams you will need to rely on other libraries.

The prior chapters showed how to manipulate data, as well as write code to make nice tables and graphs. This is useful in creating *ad-hoc* data analysis products for crime analysts, but often a large portion of your job is to create standardized reports. The subsequent chapter shows how to create a standardized report using Jupyter notebooks, which can generate documents automatically that update tables and graphs over time.

9 Creating Automated Reports using Jupyter

The last several chapters have detailed various aspects of conducting data analysis – querying tables, manipulating dataframes, and creating graphs. A common component of crime analysts jobs are to create various regular reports. When I worked as a crime analyst I had daily briefings I would send to the chief and the mayor for crimes in the prior day, a weekly command staff meeting where I would present up to date crime statistics, and a monthly CompStat like meeting that had individuals from not only the local police department but also other collaborating criminal justice agencies (like parole, probation, and the local district attorney).

Automating these reports is a very important skill set for crime analysts. The more tasks you can automate, even if not 100% automated, will make your job so much easier. If you have to spend several hours every day making a daily briefing report, figuring out code to help whittle that time down can help greatly improve your capacity as an analyst.

This chapter will show using *Jupyter notebooks* to help create automated reports.

9.1 Getting Started with Jupyter

To get started with Jupyter, create an empty folder to save your work in. Here I created a folder `B:\notebook_example` (you can however choose a different drive or location for your own work). Within this folder, I additionally created a folder `src`, and in that folder I placed a script `mpl_looks.py`, that has the content:

```
'''
Theme for graphs
'''

import matplotlib

new_theme = {'axes.grid': True,
             'axes.axisbelow': True,
             'grid.linestyle': '--',
             'legend.framealpha': 1,
             'legend.facecolor': 'white',
             'legend.shadow': True,
             'legend.fontsize': 14,
             'legend.title_fontsize': 16,
```

```
                'xtick.labelsize': 14,
                'ytick.labelsize': 14,
                'axes.labelsize': 16,
                'axes.titlesize': 20,
                'axes.titlelocation': 'left',
                'figure.dpi': 100}

matplotlib.rcParams.update(new_theme)
```

So you can see this script just has my updated graph looks.

Now that our setup for the chapter is ready, navigate your command prompt to the folder you just created, and then type:

```
jupyter notebook
```

```
Anaconda Powershell Prompt (Python)
(base) PS B:\notebook_example> jupyter notebook
```

After you hit enter, this will then launch your internet browser. (Depending on your Jupyter version, it may by default be in dark mode, instead of light mode as I have it here.) Right click on the screen, and then click the `New Notebook` option.

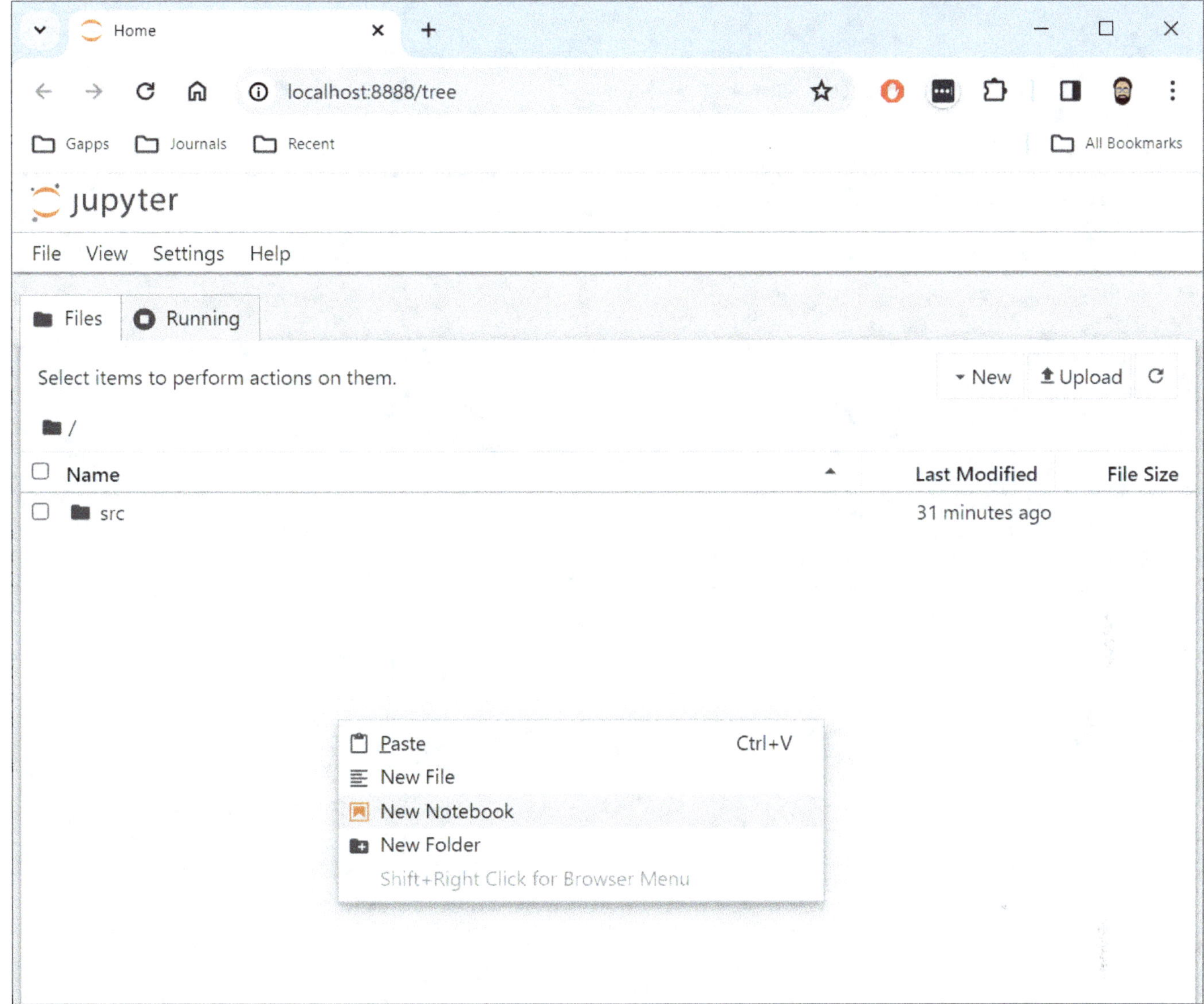

This will open up another webpage, and you may be greeted with a screen to `Select a Kernel`. Go ahead and just choose whatever default python kernel option comes up. Kernel's are similar to python environments that I talked about in Chapter 5.

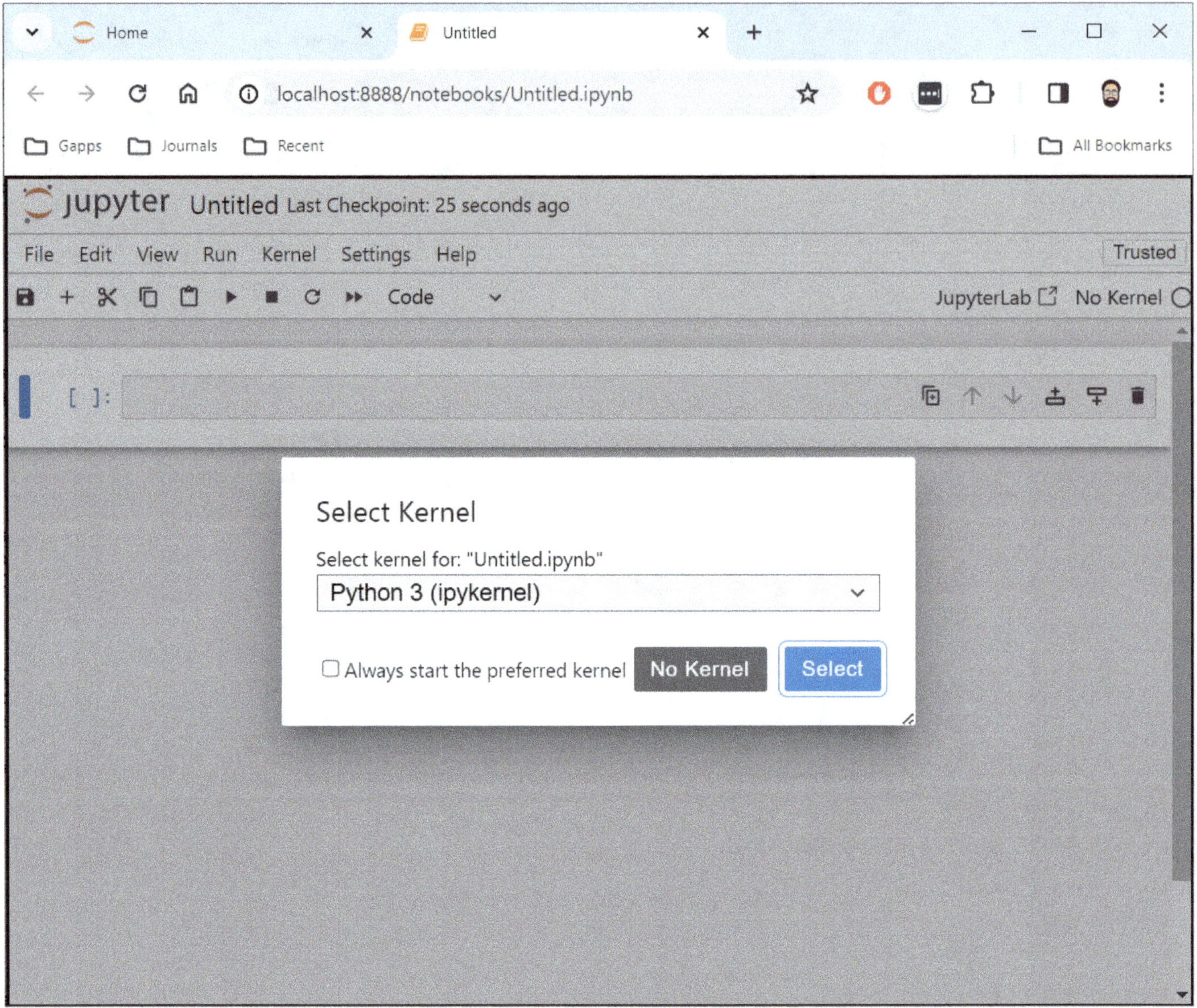

> Note
>
> When running the command `jupyter notebook`, technically what is happening is that you are creating a local *server*. You can think of this server as a little python environment, similar to the REPL, that lets you run python code. More complicated work environments can technically have the Jupyter notebook server run on different machines, and you just *access* the server from your local machine.

Before we write any code, at the top of the page, click on the `Untitled` part of the notebook. We are going to change the name of notebook to `demo.ipynb` (the file extension `ipynb` is Jupyter notebook filetype).

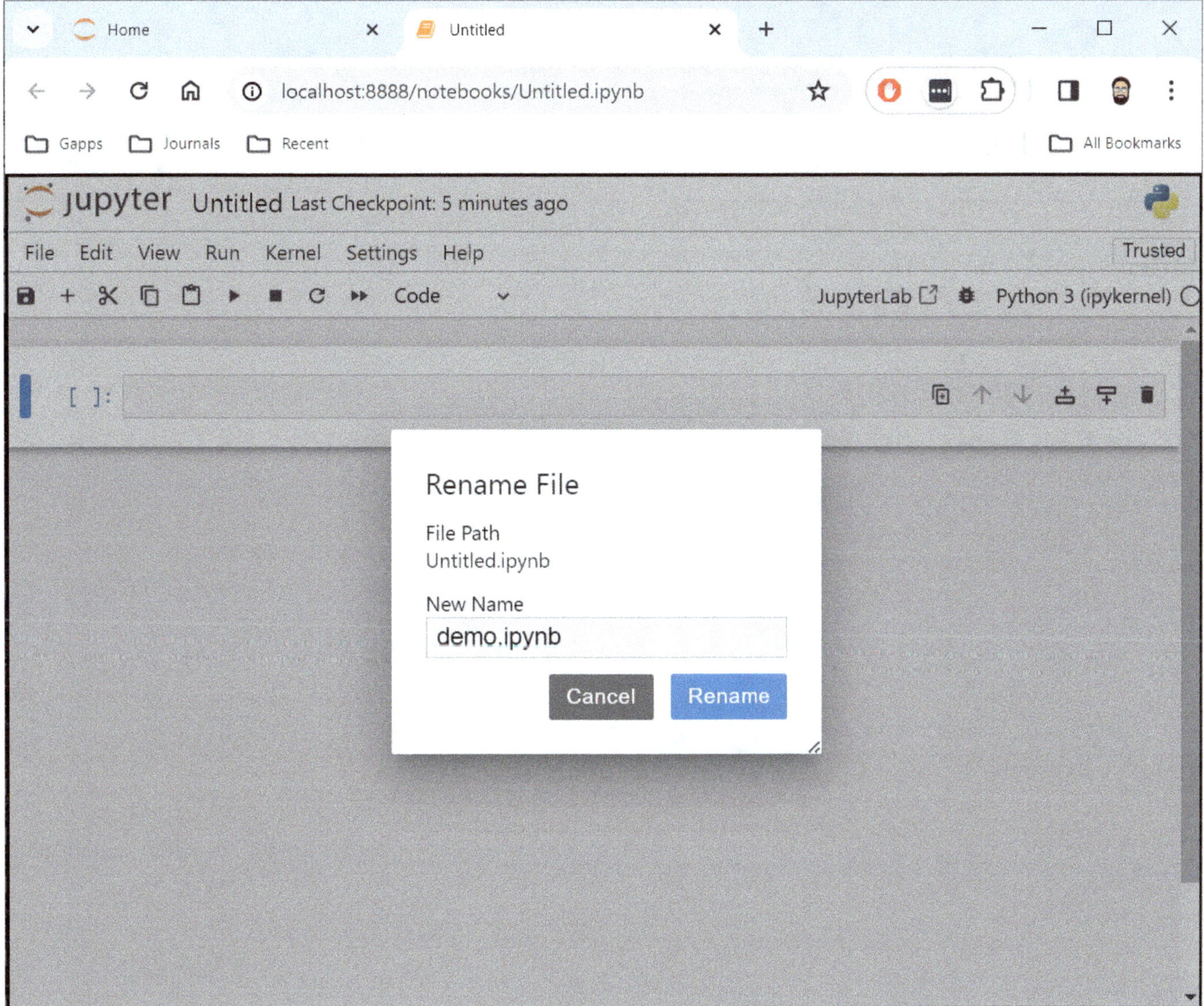

Now lets run a very simple command, type `print('hello world')` into the box:

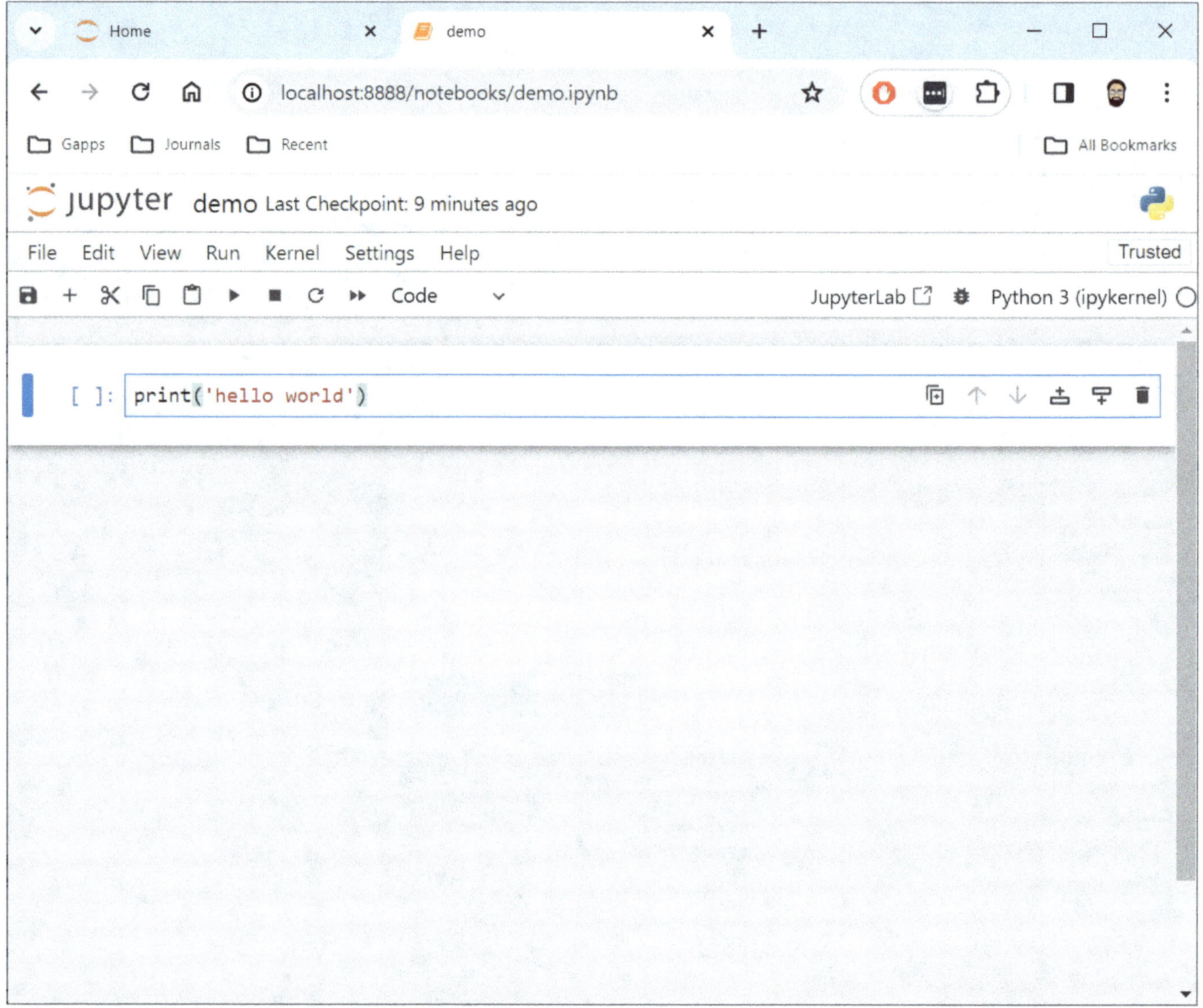

Now if you just hit the `Enter` key, this will place the cursor on a new line. It will not run the code. To run the code for a single cell, you can either hit the little play arrow in the toolbar, or hit the keys `Shift` + `Enter` at the same time to *execute* the cell. Once you have run a cell, you will then see the output of that command, as well as another new cell below the original one.

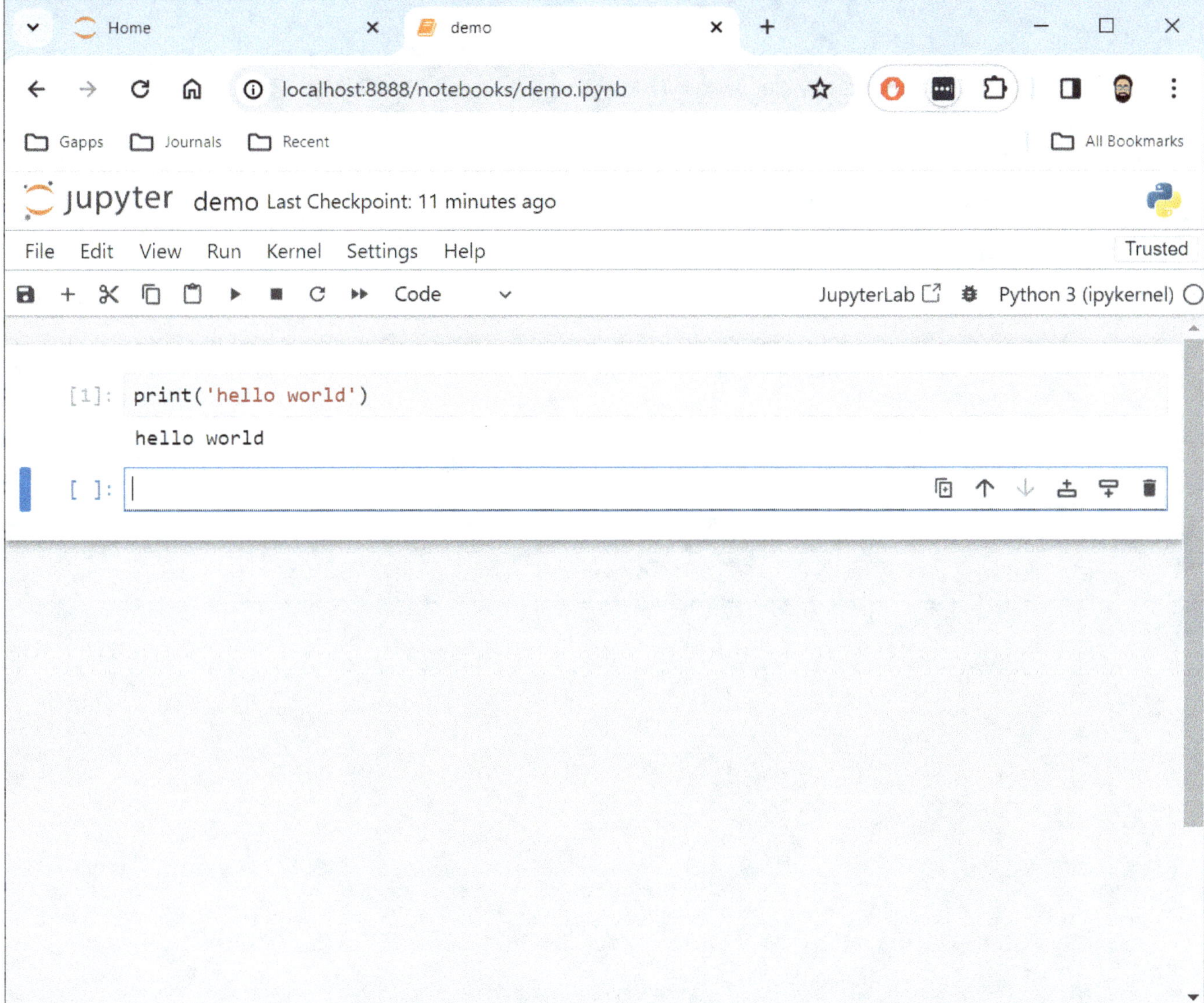

Place the cursor back at the original cell, delete the prior command, and instead write:

```
# Basic demo notebook
x = 1
print(x)
```

in the cell. Then go ahead and run that cell again.

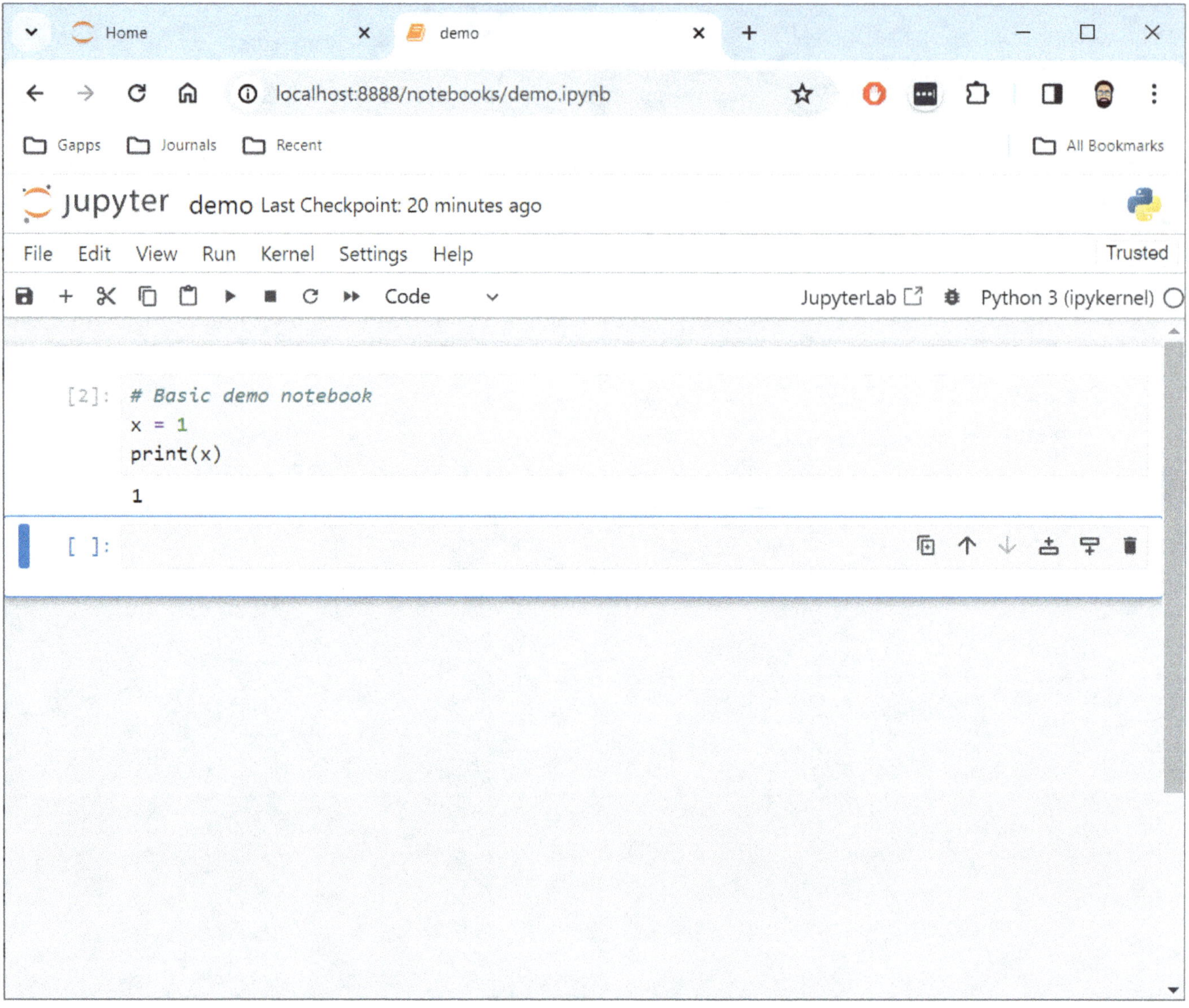

This illustrates several things. Unlike running code in the REPL, you can go back and edit code. Also you can write commands across several lines at once. You can intersperse any python commands I have shown previously.

Also note that there is a little number to the left of the cell you just ran that now says `2`. (If you go back to the prior `hello world` screenshot, you will see that it lists `1`). Jupyter notebooks have a global state, so elements defined in a prior cell can be accessed in a subsequent cell. In the second cell, now run `y = x + 1` and then `print(y)`.

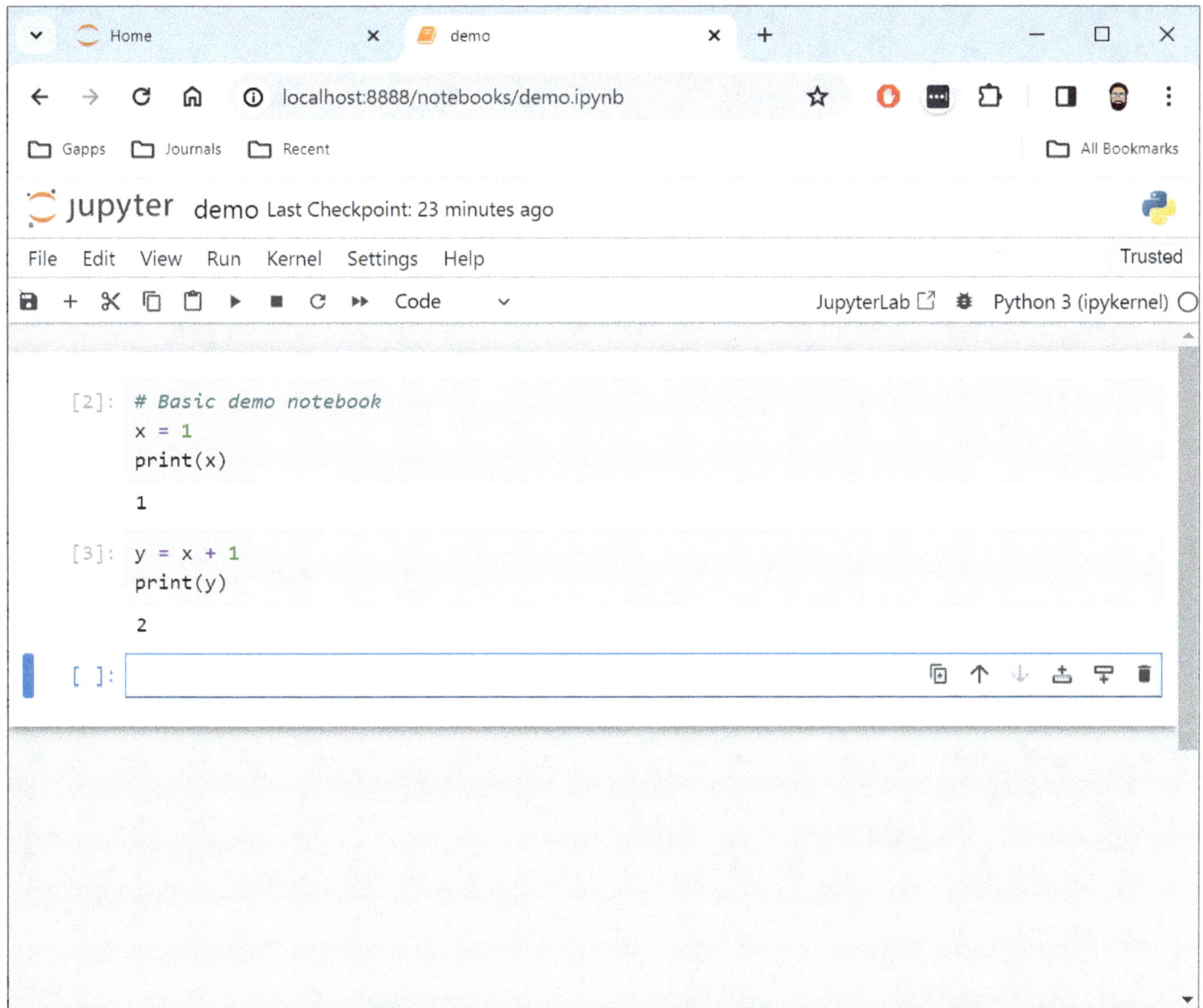

One thing to watch out for in cells, is that they do not force you to run code in the correct order. Edit the first cell to include a line that has `z = y + x`, and then hit `Shift` + `Enter`:

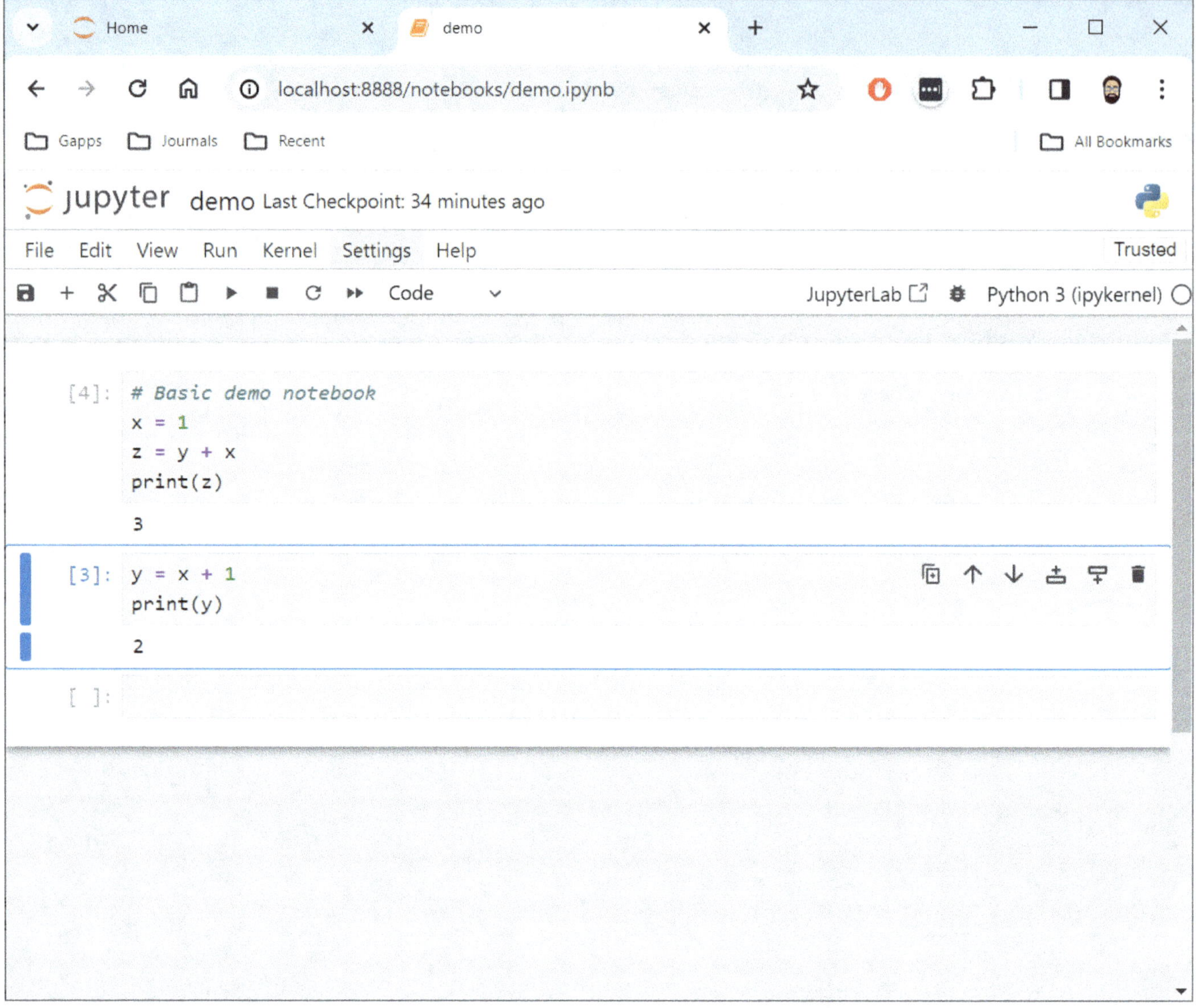

You will see that the first cell has a number 3, and runs without error. This is a common mistake when writing code in notebooks, that when writing interactive code you execute parts out of order.

To make sure this does not happen, it is common once your code is finished, to run the entire notebook at once. In the toolbar, select the Run tab and then hit the option `Restart Kernel and Run All cells`.

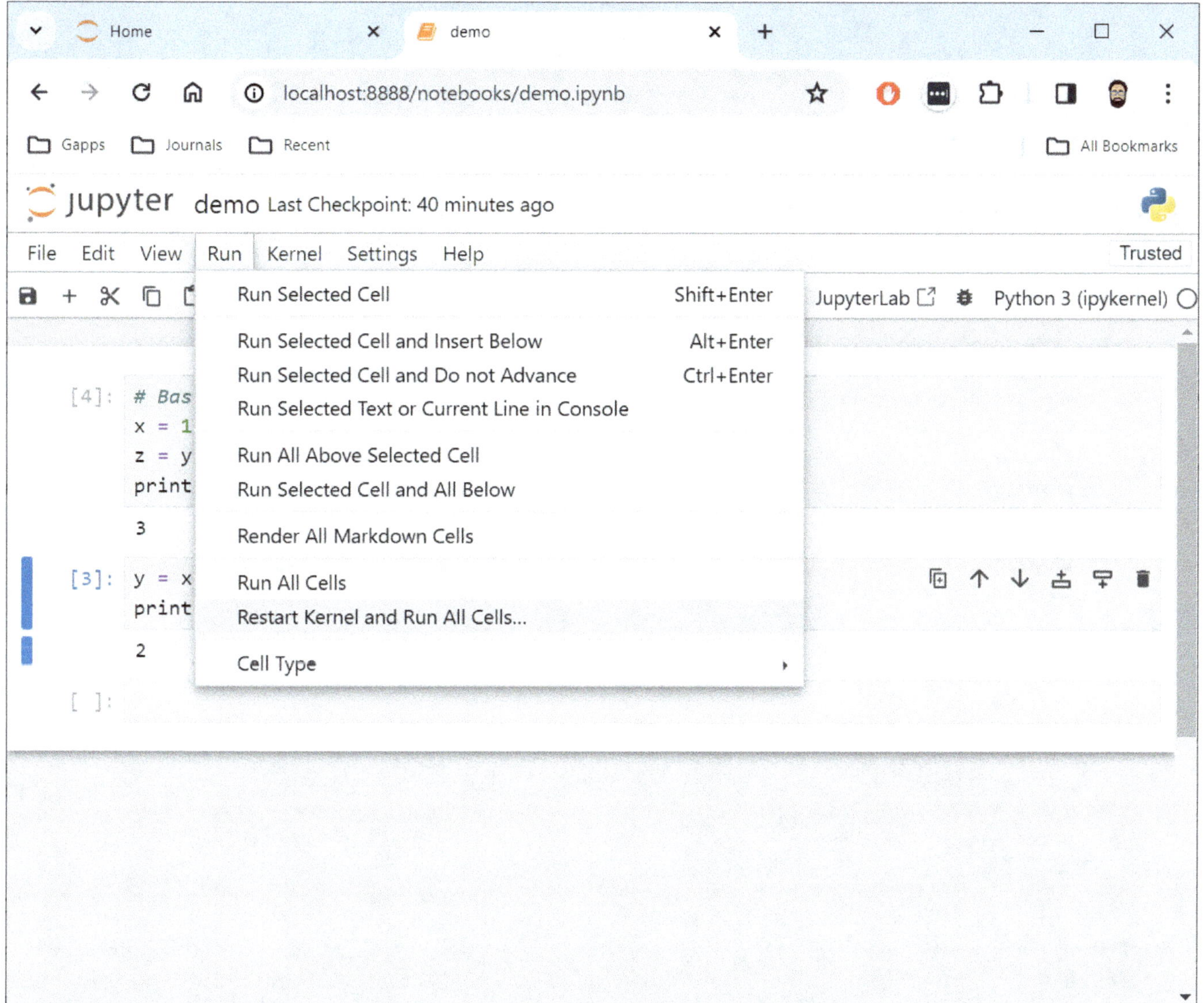

You will get a warning about restarting the kernel and losing all variables. This is ok here, and so click the `Restart` button.

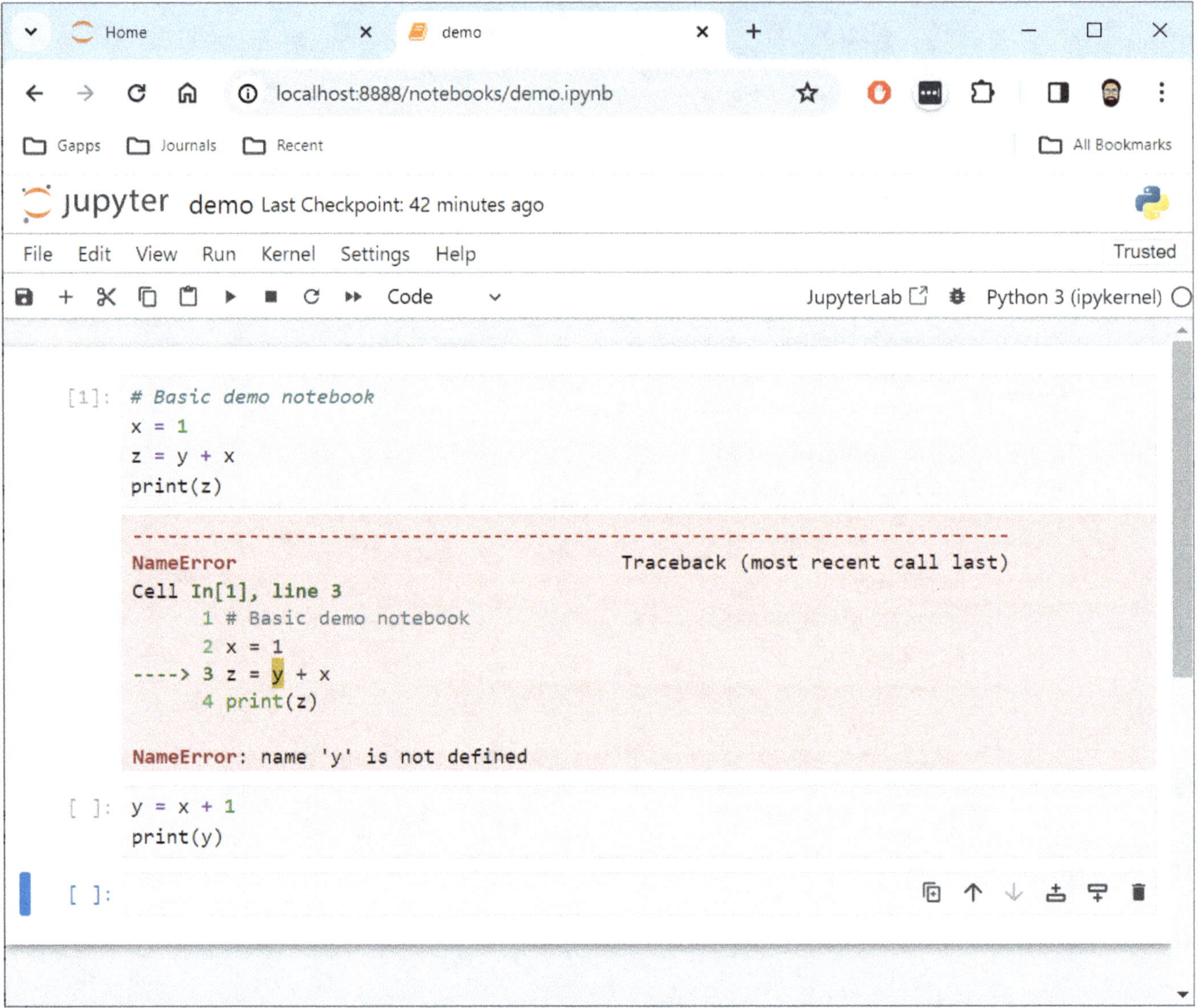

You can see now that there is an error that `y is not defined`, and that the second cell is not even run. Go ahead and edit it so that the `z` variable is in the second cell after you have defined `y`, and then Restart the kernel and re-run the entire notebook again.

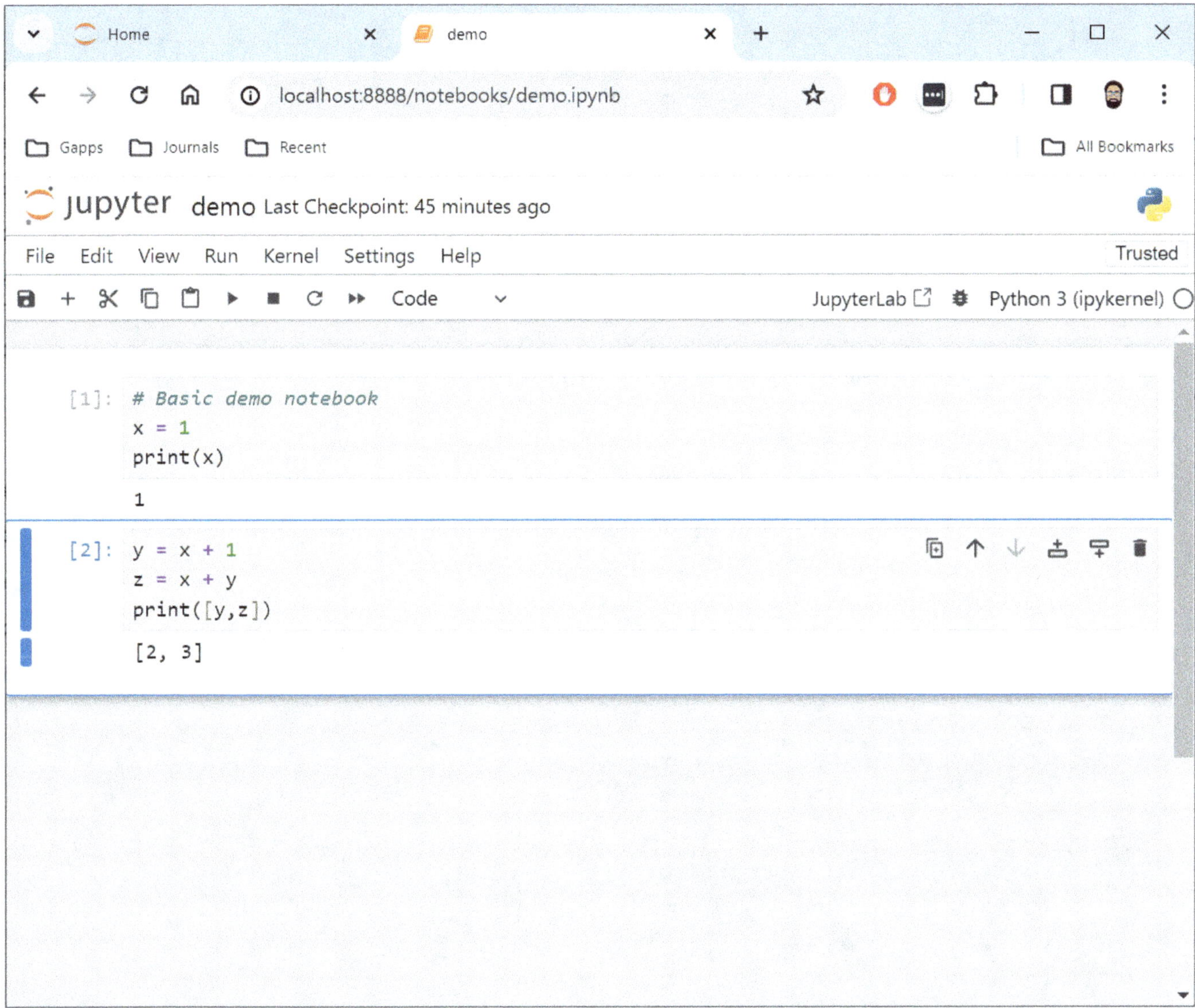

Now the code runs correctly.

9.2 Cell Basics

So far I have shown only python code cells in this notebook. You can have other types of cells though as well. The most common example is a *markdown* cell. This is a cell type that does not run code, but has other information. In your first cell, click the icon on the top right that has the tooltip `Insert a Cell Above`.

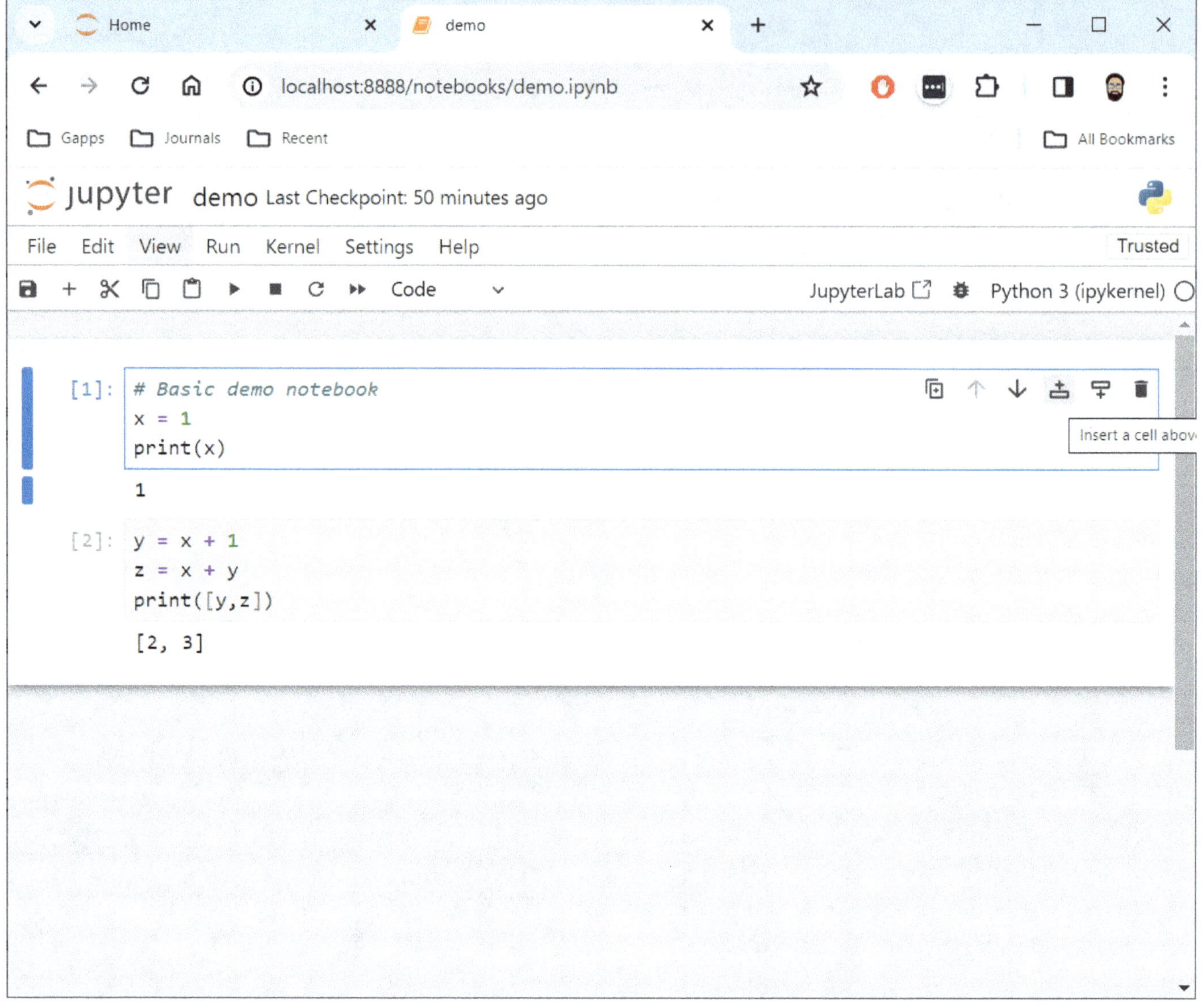

Then in the toolbar, edit the dropdown that by default says `Code` to `Markdown`.

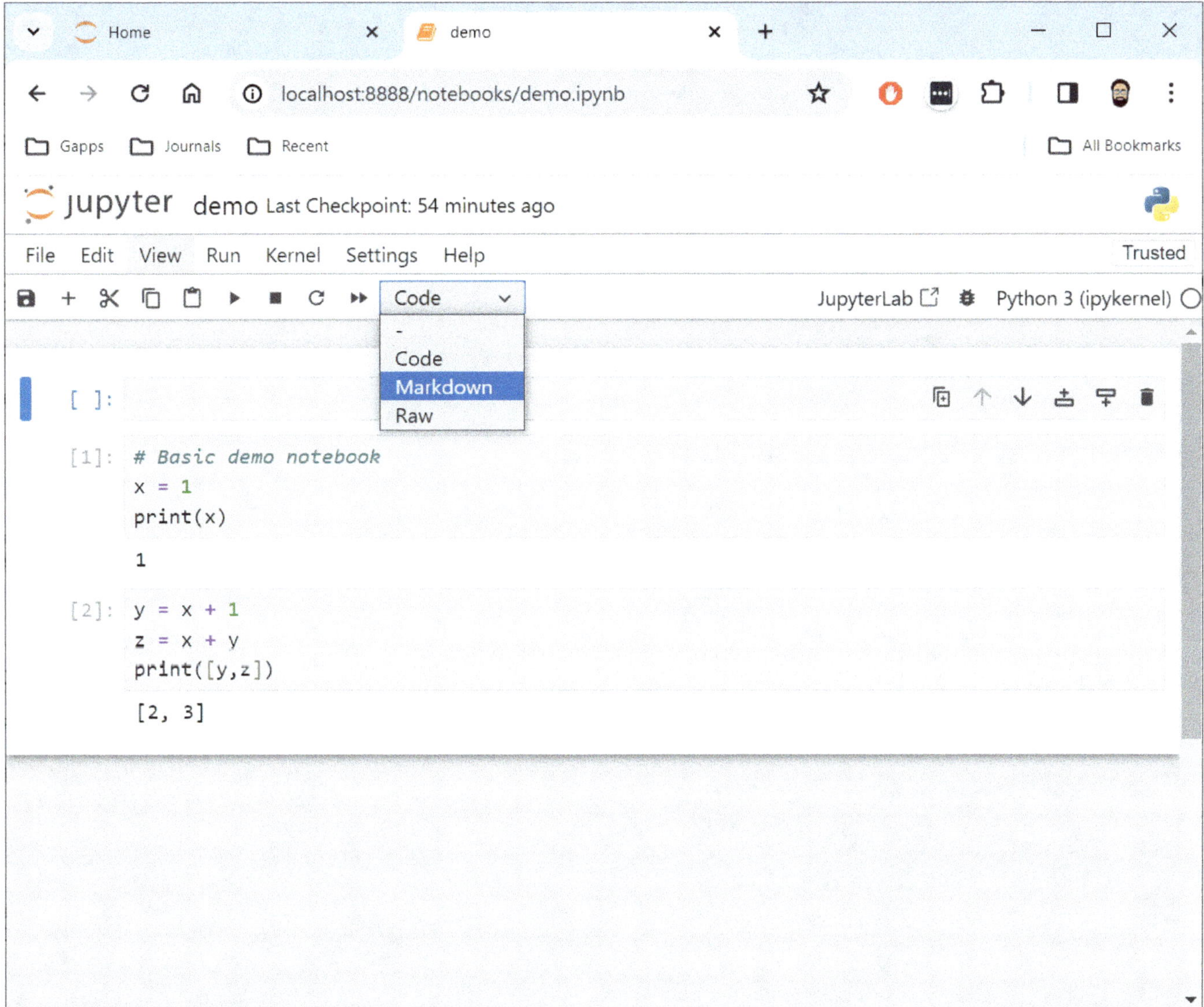

Now in the markdown cell, write the code

```
# Title

This is a *demonstration* notebook. It shows:

 - how to execute cells
 - how to create markdown cells
 - how cells can be out of order
```

Before you execute that cell, your notebook will look like this:

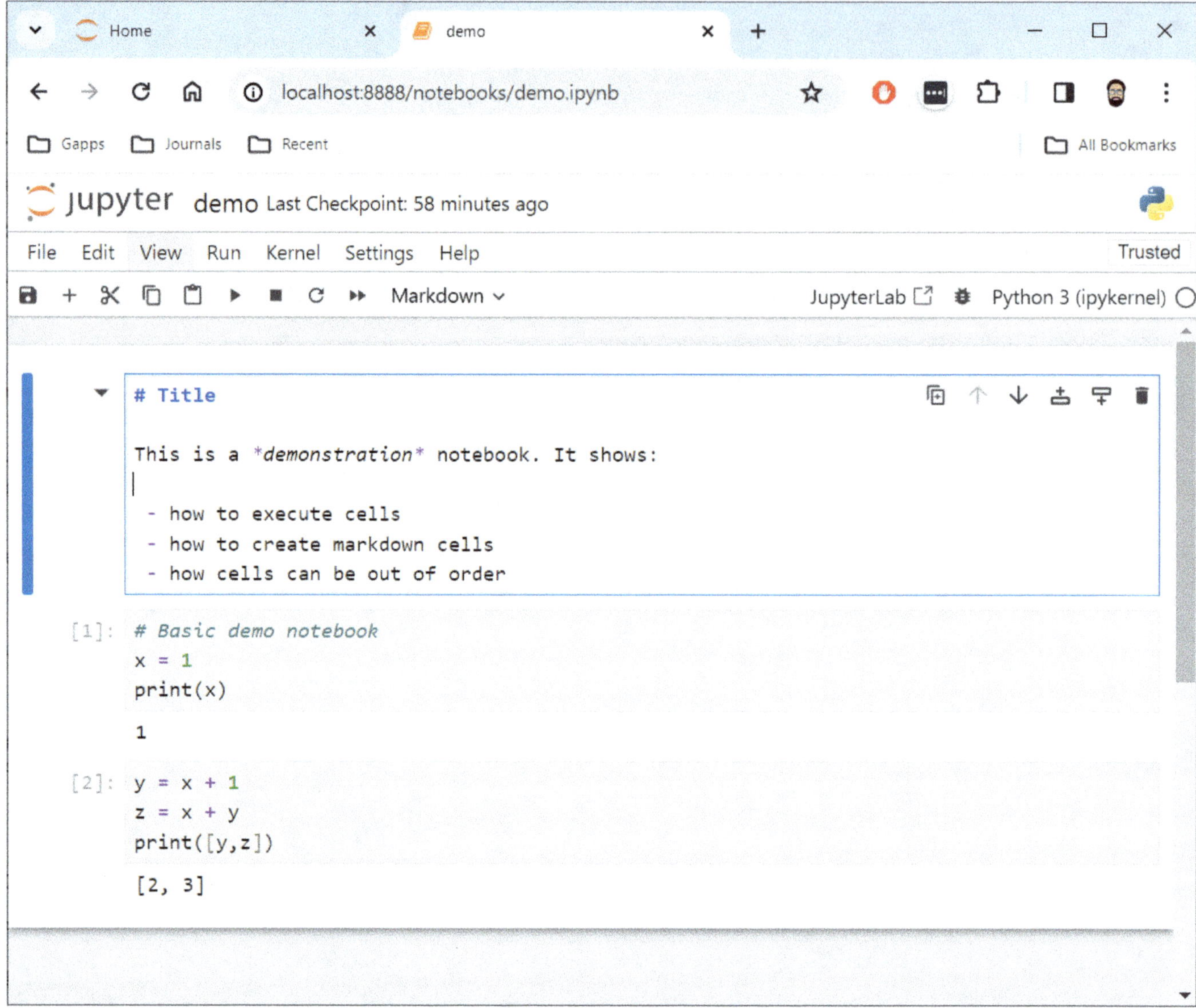

When you execute this cell, it will then render the markdown.

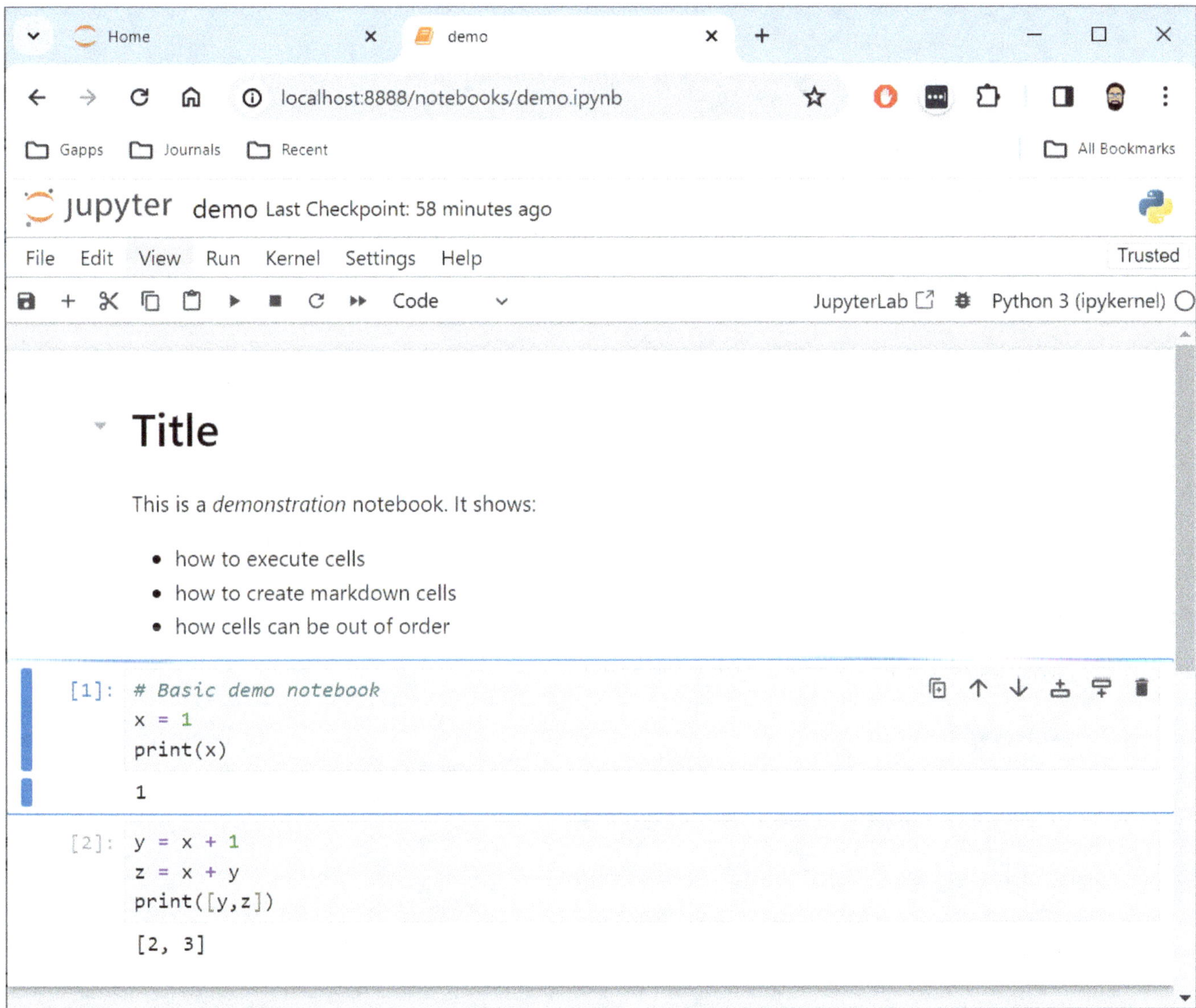

Markdown is a way to use plain text to write content. For example, to make italics in a word document, you would use the GUI interface, or select some text and then hit `Ctrl` + `i` to change to italics. This is called a *what you see is what you get* word editor. In markdown, you would write text enclosed in two asterisks, e.g. `*text*`, and then whatever engine renders the markdown to a final document at that stage creates the text as italics.

Using markdown cells is convenient way to have descriptive text, such as different document sections, inside of a single notebook. Here I created two additional markdown cells to signify different subsections of the notebook, as well as highlight a few different ways to write markdown code.

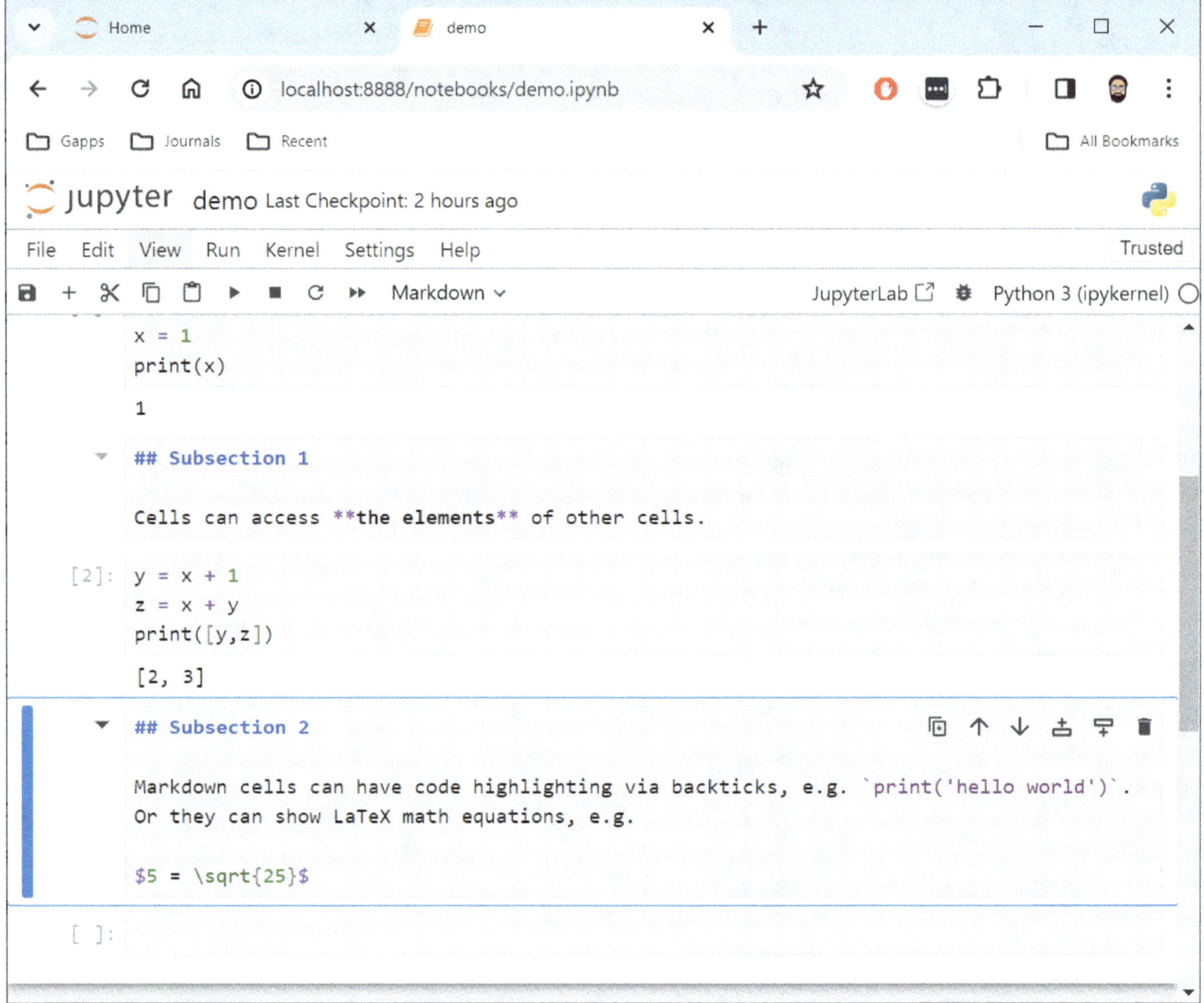

> **Note**
>
> There are many different things you can do in markdown, and there are different types of markdown as well. For example, some markdown rendering engines you can use underscores for italics or bold, e.g. `_emphtext_` instead of `**emphtext**`. Another common one is to create hyperlinks via `[text](url)`. For example, at the beginning of your document you could have a line `[Andrew Wheeler](mailto:andrew.wheeler@crimede-coder.com)`.

When rendered, this looks like:

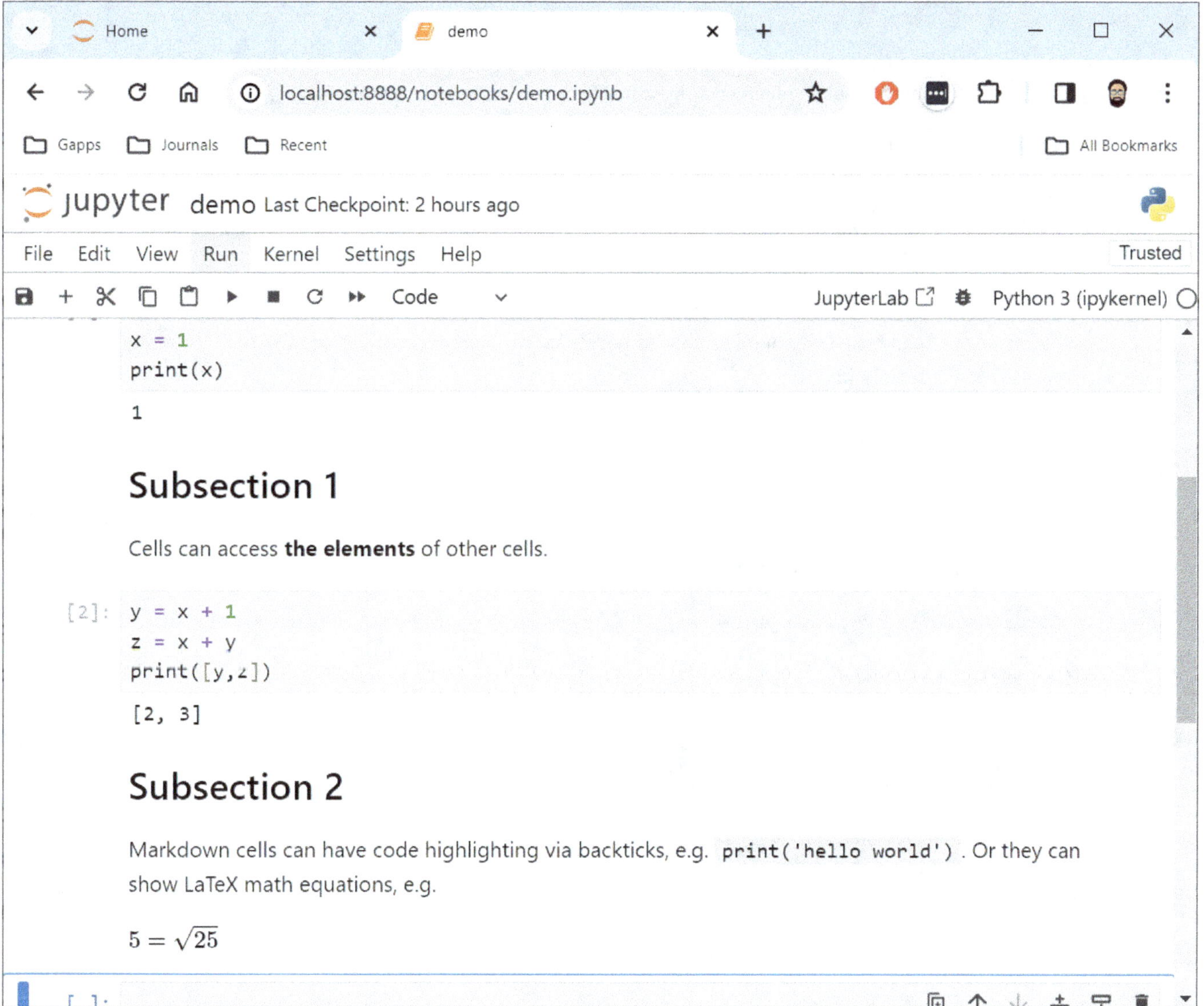

The final example I want to show is using cell aliases. Sometimes you want to run *shell* commands. You can do this by starting a python line with an exclamation point. For example, `!dir` will show the current directory (on Windows, if on Mac or Unix, use `!ls`).

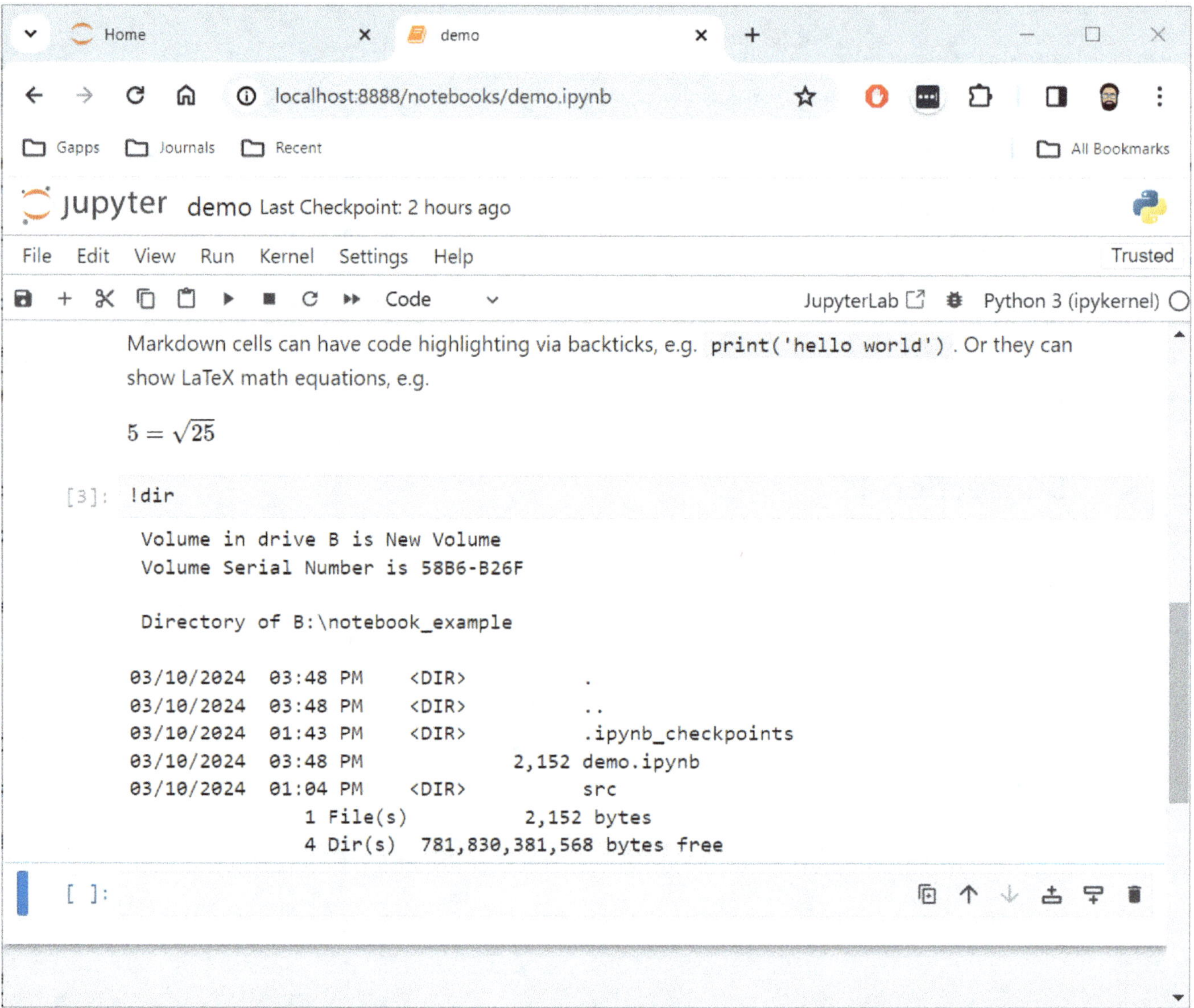

This is called a *line* magic command.

> **Note**
>
> Sometimes you will see notebooks have magic lines like `!pip install package1 package2`. This is in general not good practice, for reasons discussed in Chapter 5 on the need to make new python environments when installing packages. If you need to install new packages, in general you should create a new python environment and a new *Jupyter kernel*, and then have the notebook associated with that new kernel. From the command line, this would look like:

```
conda create --name new_env python=3.10
conda activate new_env
pip install -r requirements.txt
ipython kernel install --name "new_kernel" --user
```

The last example type of cell I want to show is a cell alias. This is also often used to run shell commands. You start a cell this way by first writing two percentage signs, and then the cell type. So to run a series of commands in the Windows command prompt, you would run something like below. Here I created an environment variable `JUPY_SECRET` just for illustration. And you can see that I have shell commands spread across several lines:

```
%%script cmd
pwd
echo %JUPY_SECRET%
```

And when run the results look like below:

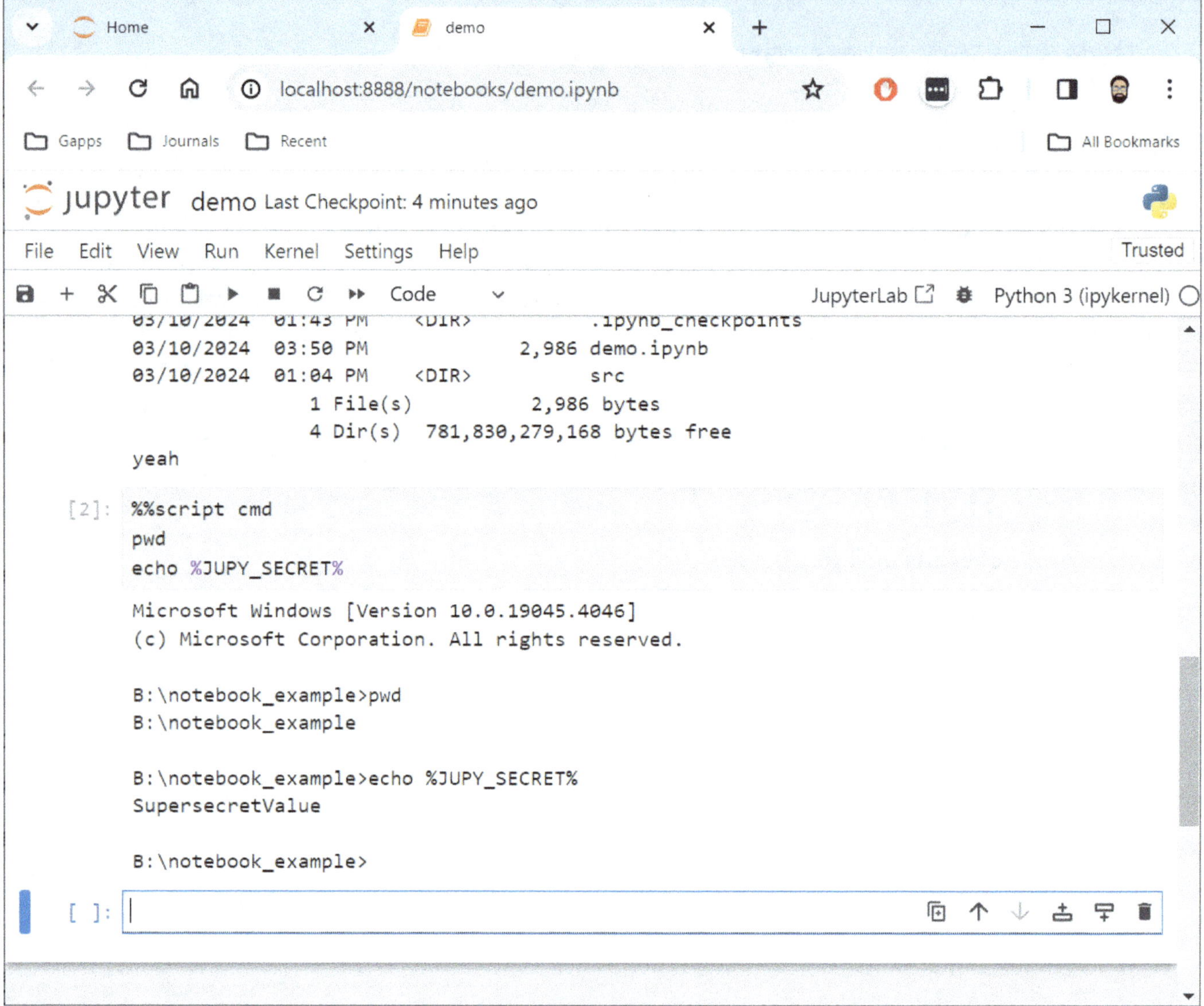

> **Note**
>
> If the `pwd` command is not available, then typing `cd` instead will accomplish the same thing, echoing the current location the script. The `%JUPY_SECRET%` environment variable will only print out something if you created that variable. To swap in another environment variable you are sure to have, you can type inf `echo %PATH%`, and it will show (likely many) locations that are on your path variable.

On Windows you could also run `%%script powershell`. On Mac or Unix you would often use `%%script bash`.

Note

You can also call shell commands directly from python. For example, say you wanted to rename a file, you could do `os.system('ren OrigFile.pdf New.pdf')`. There are not many examples in my experience where you would call system commands directly though (you can also use the command `os.rename` directly for this example).

There are other types of magic cell aliases – many different computing environments offer different types of cells. For example, Databricks or Apache Zeppelin have notebooks that allow you to use sql cells directly. More recent versions of Jupyter notebooks allow you to have javascript cells (notebooks are by default rendered as HTML in the browser). You can also extend the local Jupyter notebook environment to have cells to run different programs, such as having Jupyter notebooks run R or Stata code.

9.3 Tables and Graphs

In notebook cells, you can import libraries and functions the same as in other code. Here I import pandas, matplotlib, and a custom set of code I noted at the beginning of the chapter, the `mpl_looks` script.

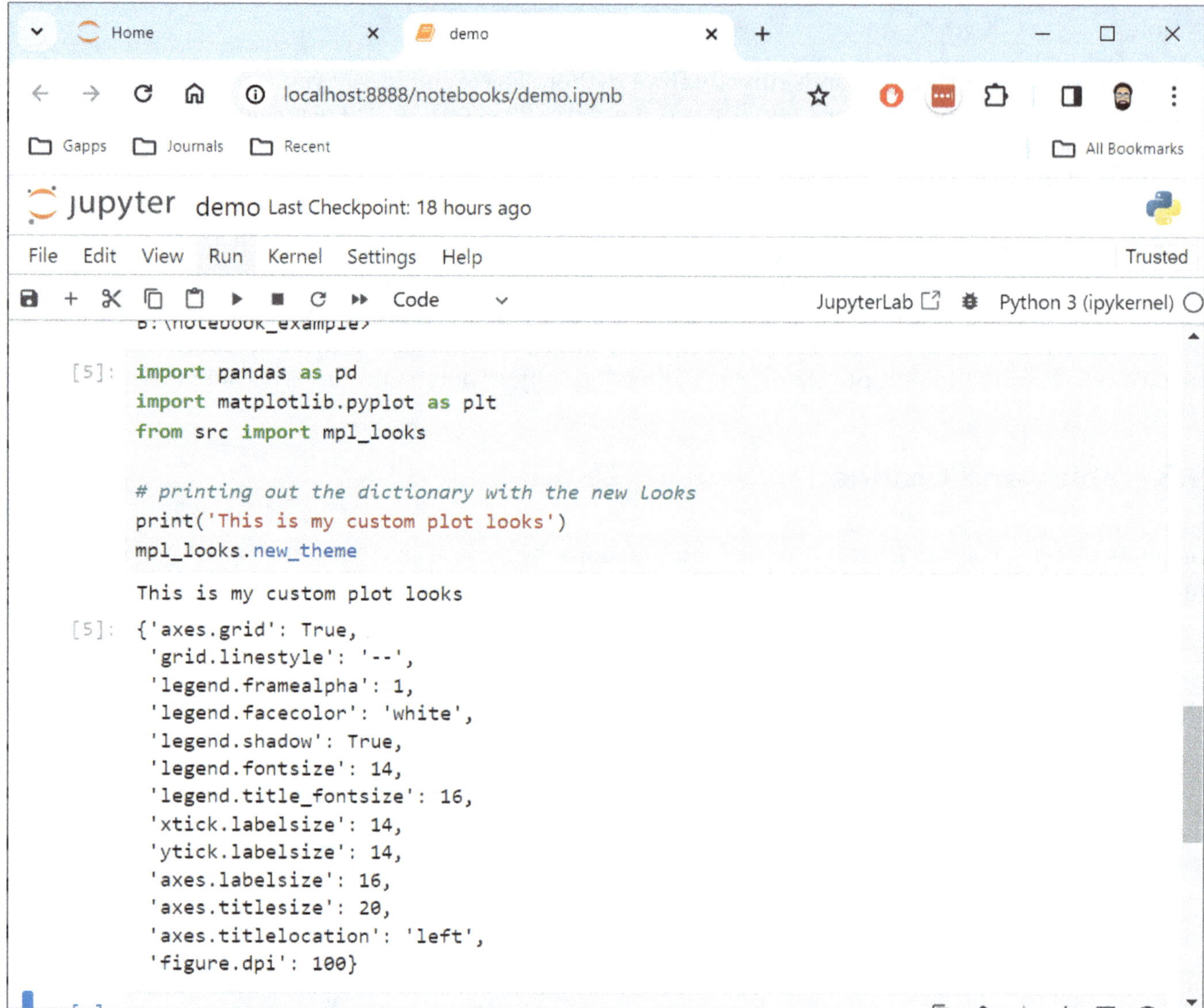

I have shown examples so far of running `print(x)`, here it is somewhat different. I have a print statement, and then on the final line of the cell just list the object, here a dictionary. So you can have a single cell output multiple items. Most objects when listed on the final line of a notebook cell like this will not look much different than if you did `print(mpl_looks.new_theme)`. But some objects, like pandas dataframes, have a nicer printing method directly to HTML tables.

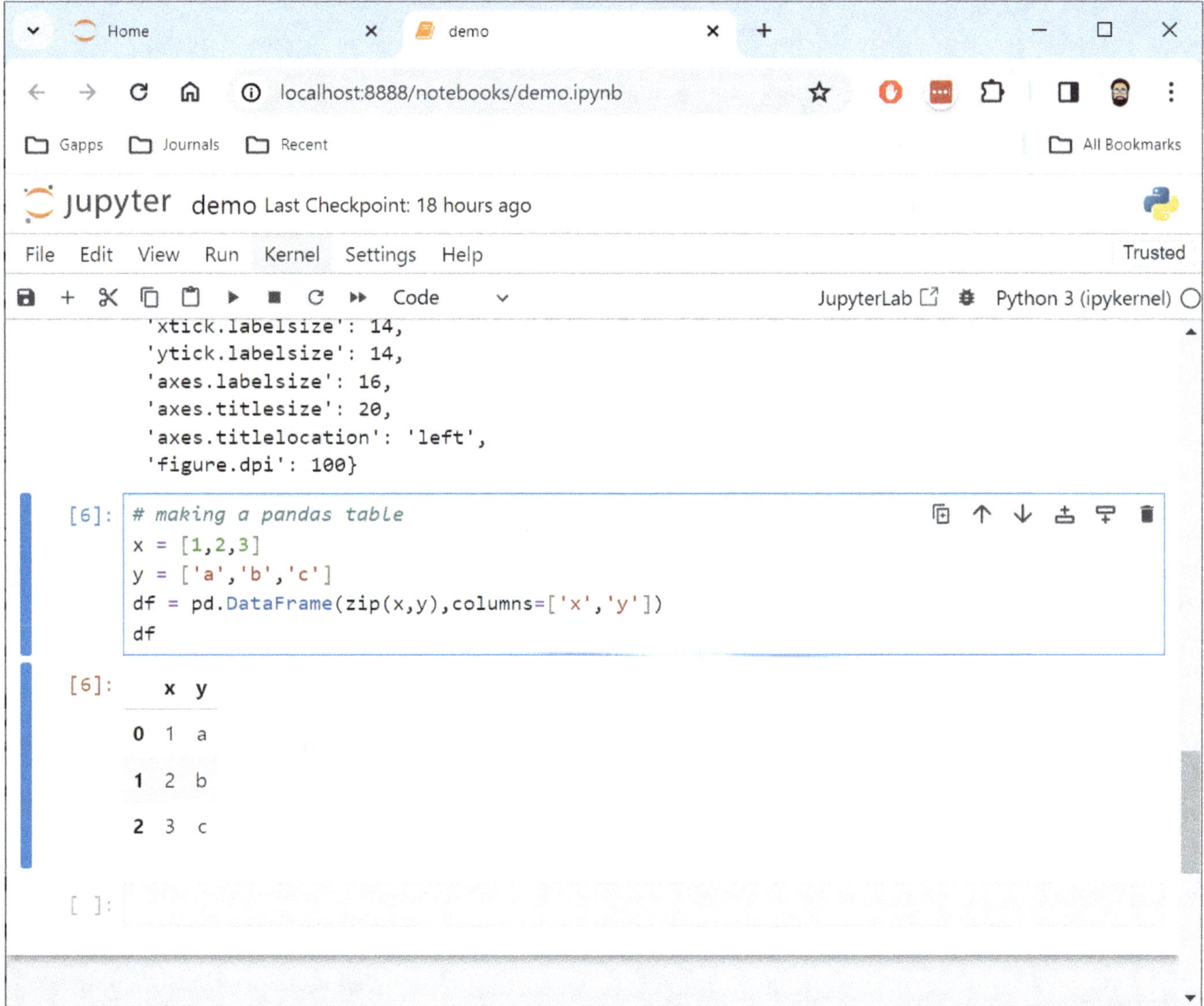

When creating reports, I often don't want to show the table index, and this example shows having the table title above in a markdown cell.

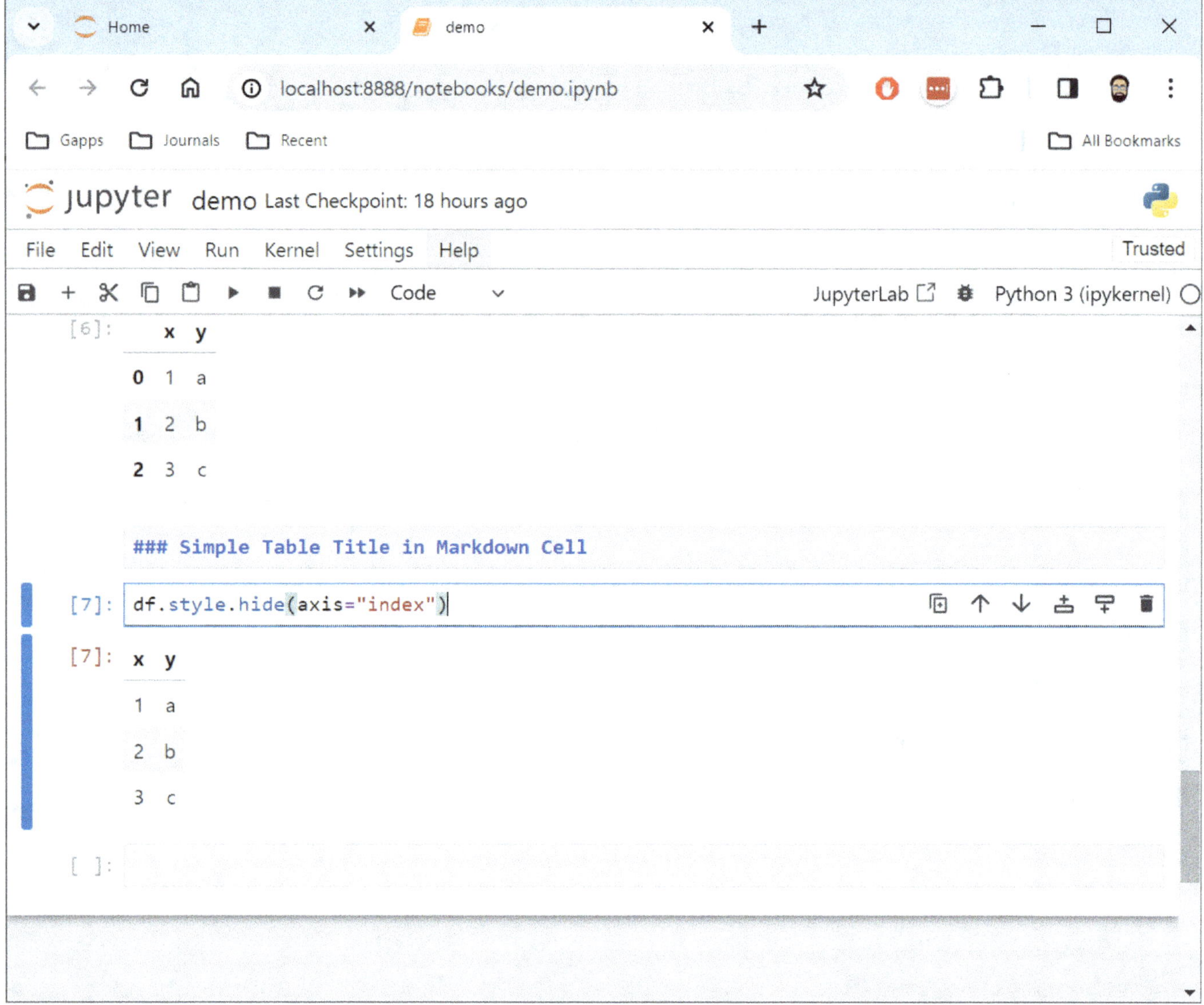

And when rendered it will look like this below.

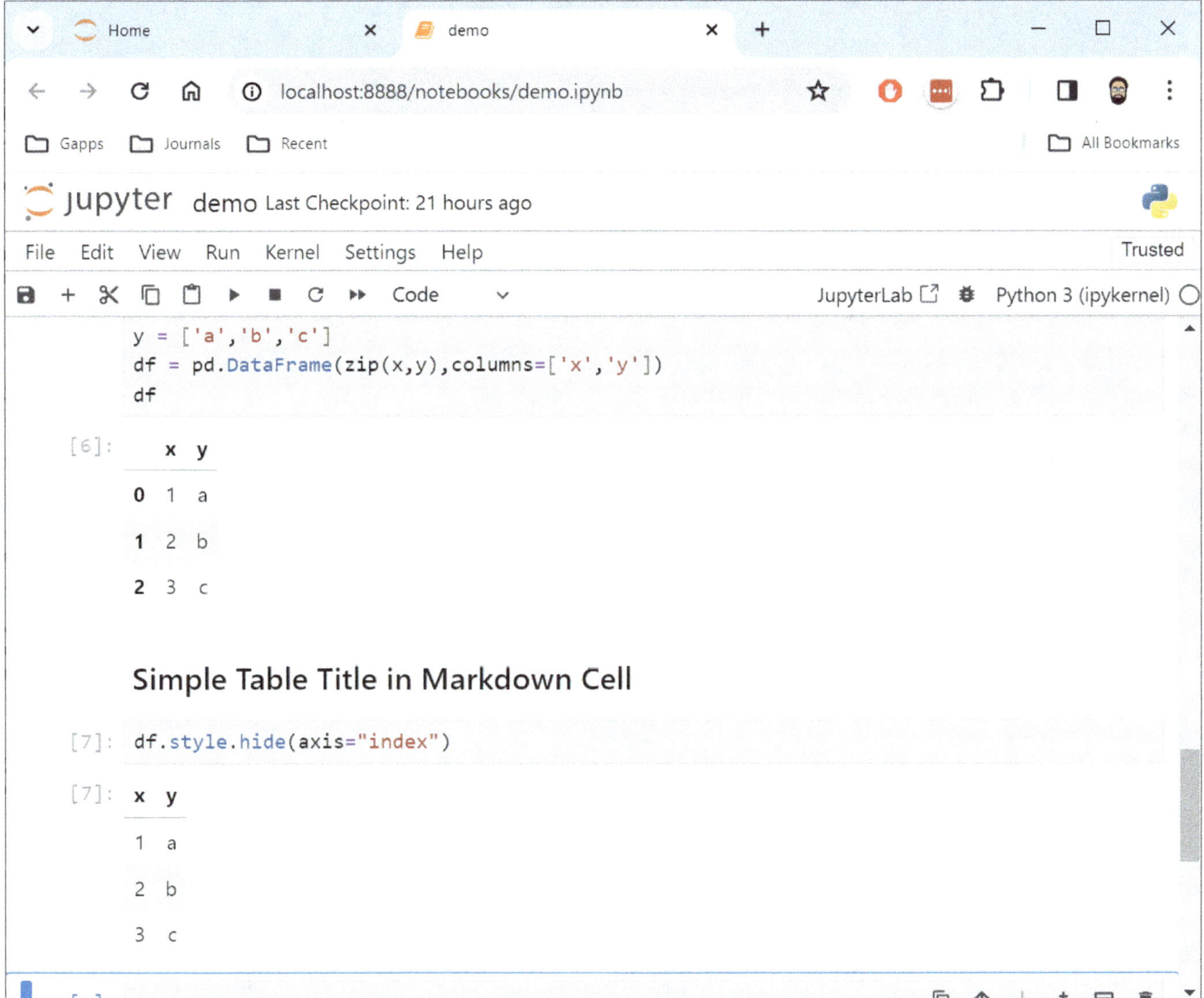

For simple tables this is sufficient, but one limitation of this is that the Markdown cell displaying the table is that the title cannot dynamically insert values. So say you wanted to put the current date in the table title, you can't do that if you have a fixed markdown cell. An alternative approach to do that is to generate markdown directly in the python code, and then have that output as the final example. This shows a function, `Markdown`, in the `Ipython.display` library to accomplish this.

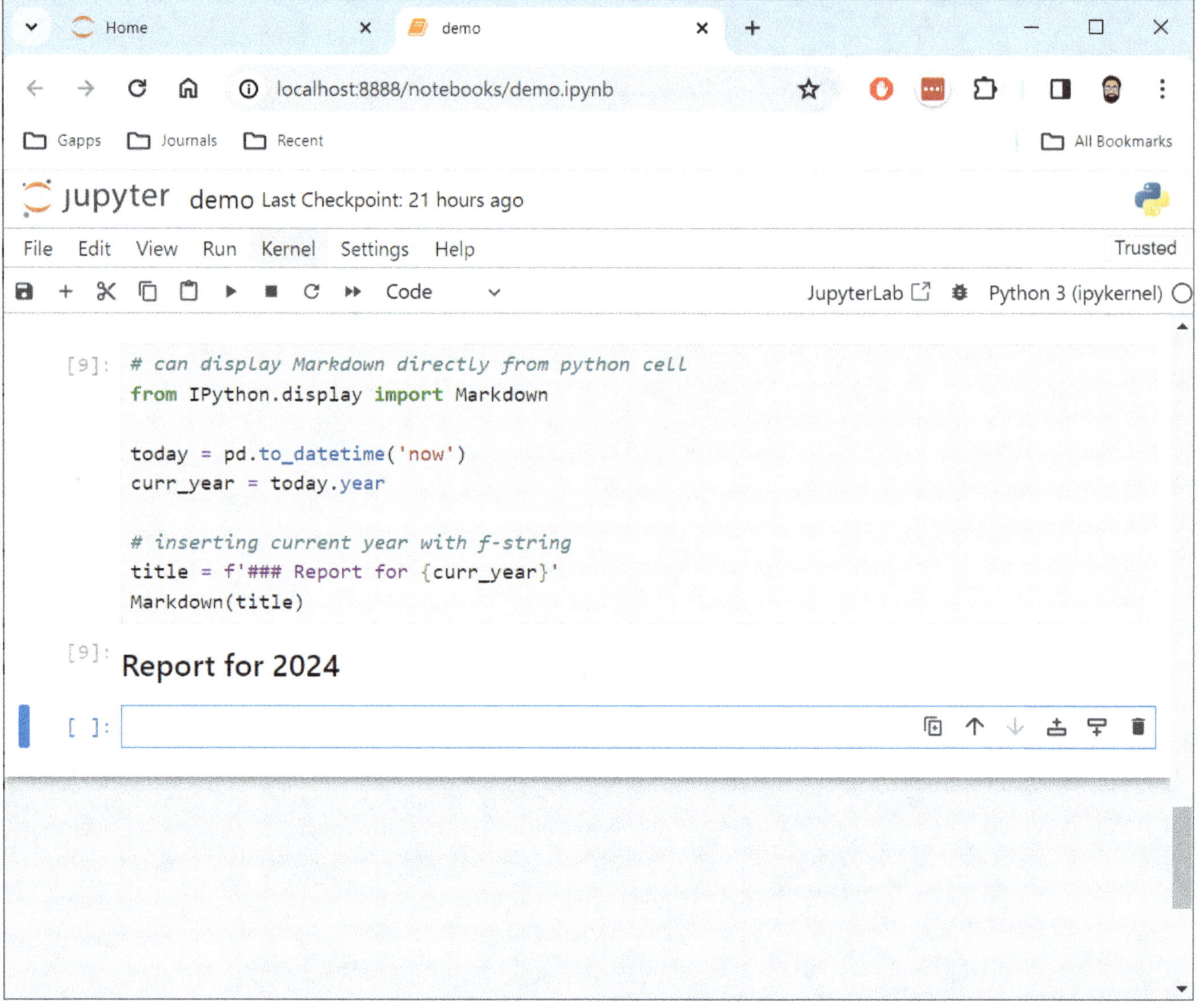

For a few different pieces of advice for tables, you should limit the number of numerical values displayed. So for example, instead of showing `0.33333`, it is better to show `0.33`. For larger numbers, using a thousands separator may be appropriate, e.g. instead of `200000` show `200,000`. Numbers should also be right aligned (which is the default in Jupyter notebooks), but text that is meant to be read you often want it to be right aligned.

Here is a table with a few extra columns to illustrate how the notebook defaults are not very good.

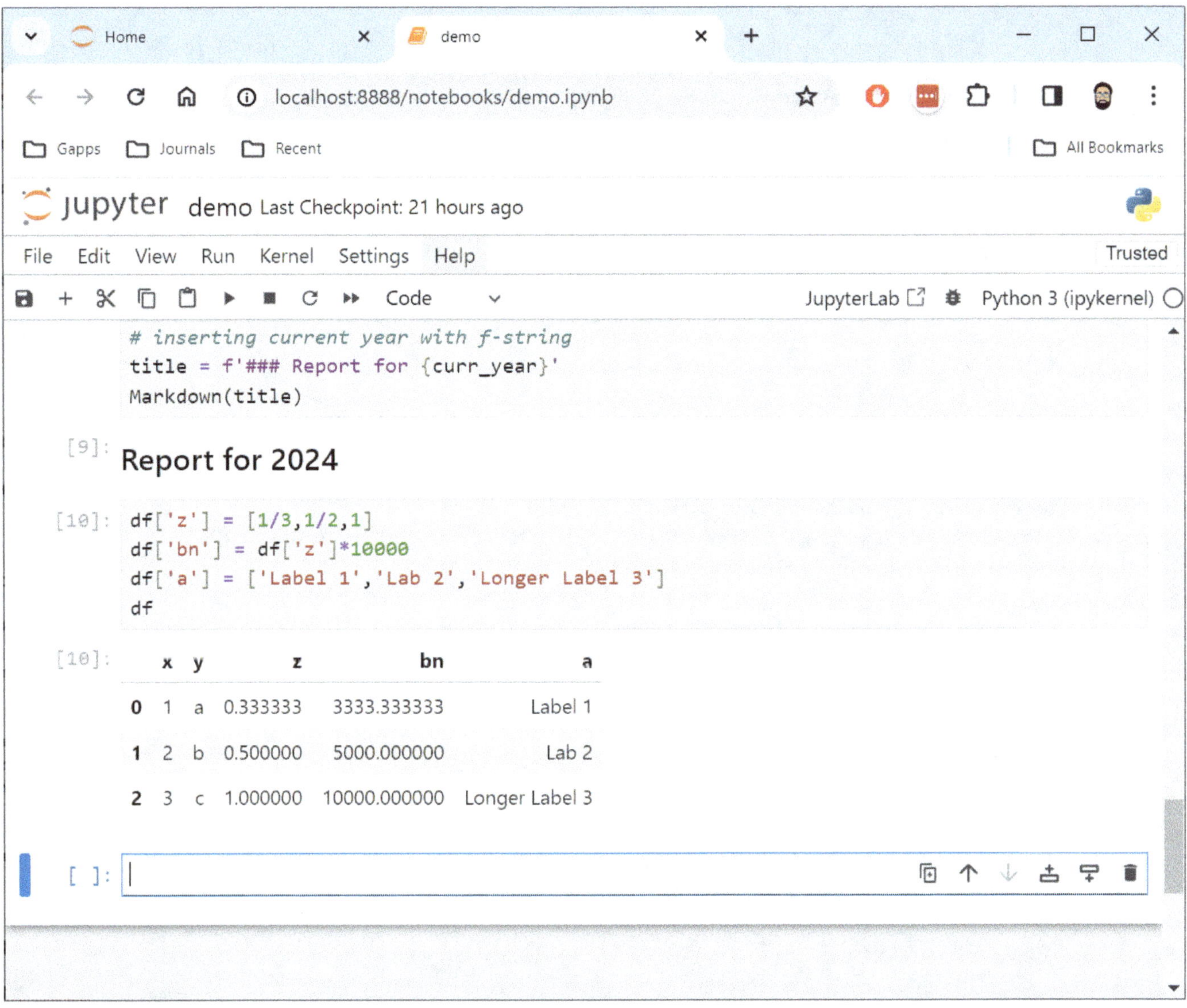

My favorite way to modify the table is using the method to convert pandas tables to markdown, `df.to_markdown`. This allows you to set the formats for the fields and column alignment. See Chapter 3 for the different format modifiers when converting numbers into strings. You can chain together multiple markdown strings. So you shows how to display the title and table together as the output of a single cell execution.

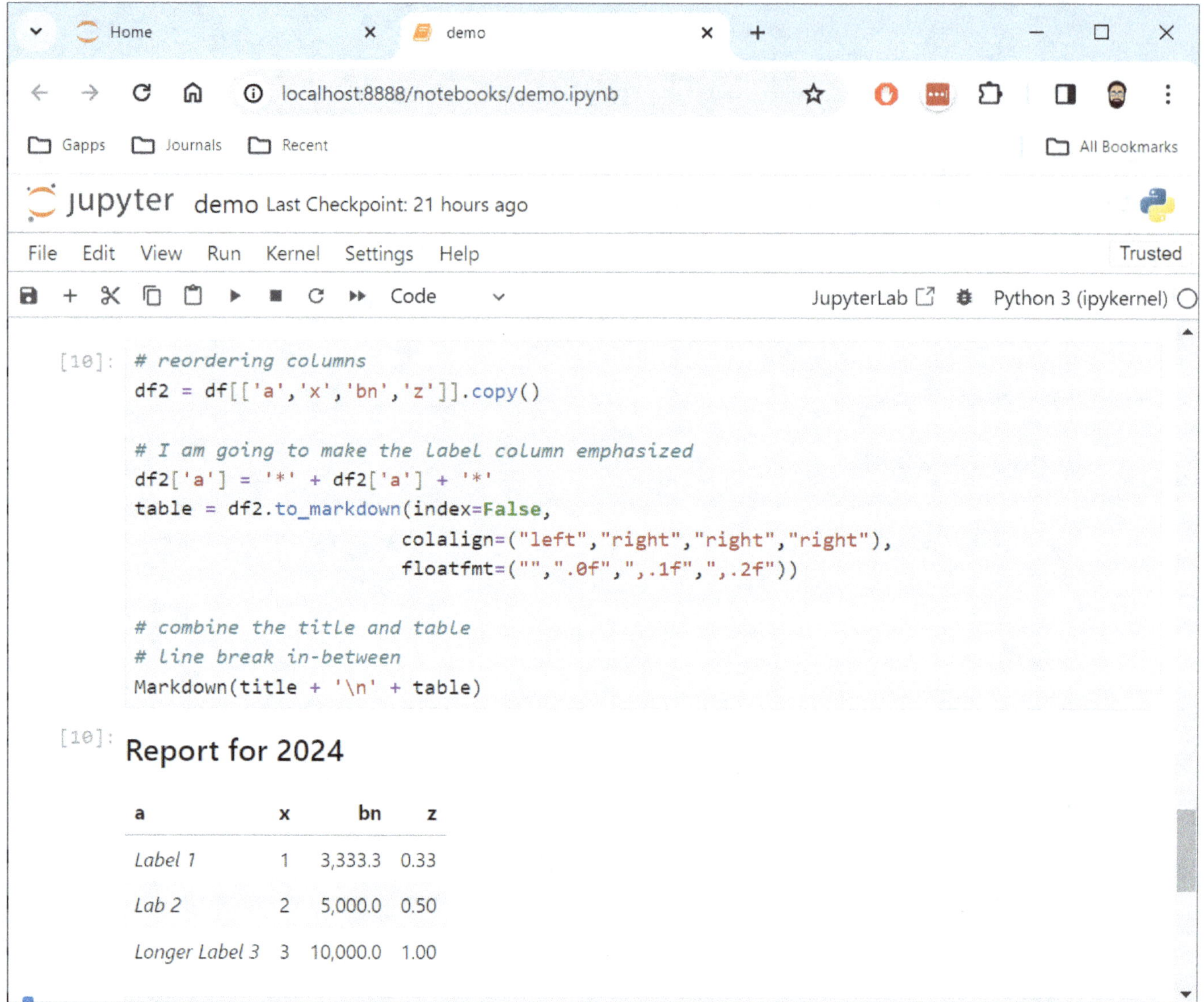

> **Note**
>
> The pandas methods `df.to_markdown` requires an outside package, *tabulate*. If you run your code and get an error message that says something like "pandas dataframe has no method to_markdown", you will need to install the tabulate package, e.g. `pip install tabulate` in your current python environment.

There are many other methods to manipulate tables to get them to look a particular way. The other most common way is instead of converting tables to markdown, convert them directly to html, e.g. `df.to_html`, and then use `from IPython.display import HTML` the same as the `Markdown` function. If you wanted to highlight a specific row as a specific color for example, I would use this method. You can also convert an entire table to string formats, and then manipulate those strings how you want them to look. But the markdown or HTML approach tends to be easier to automate and cover multiple scenarios.

The final examples I wanted to show in this section are graphs. Graphs in Jupyter notebooks are similar to tables, in that you can generate the graph output as the final output for a cell. Here I use the pandas method to generate a bar graph.

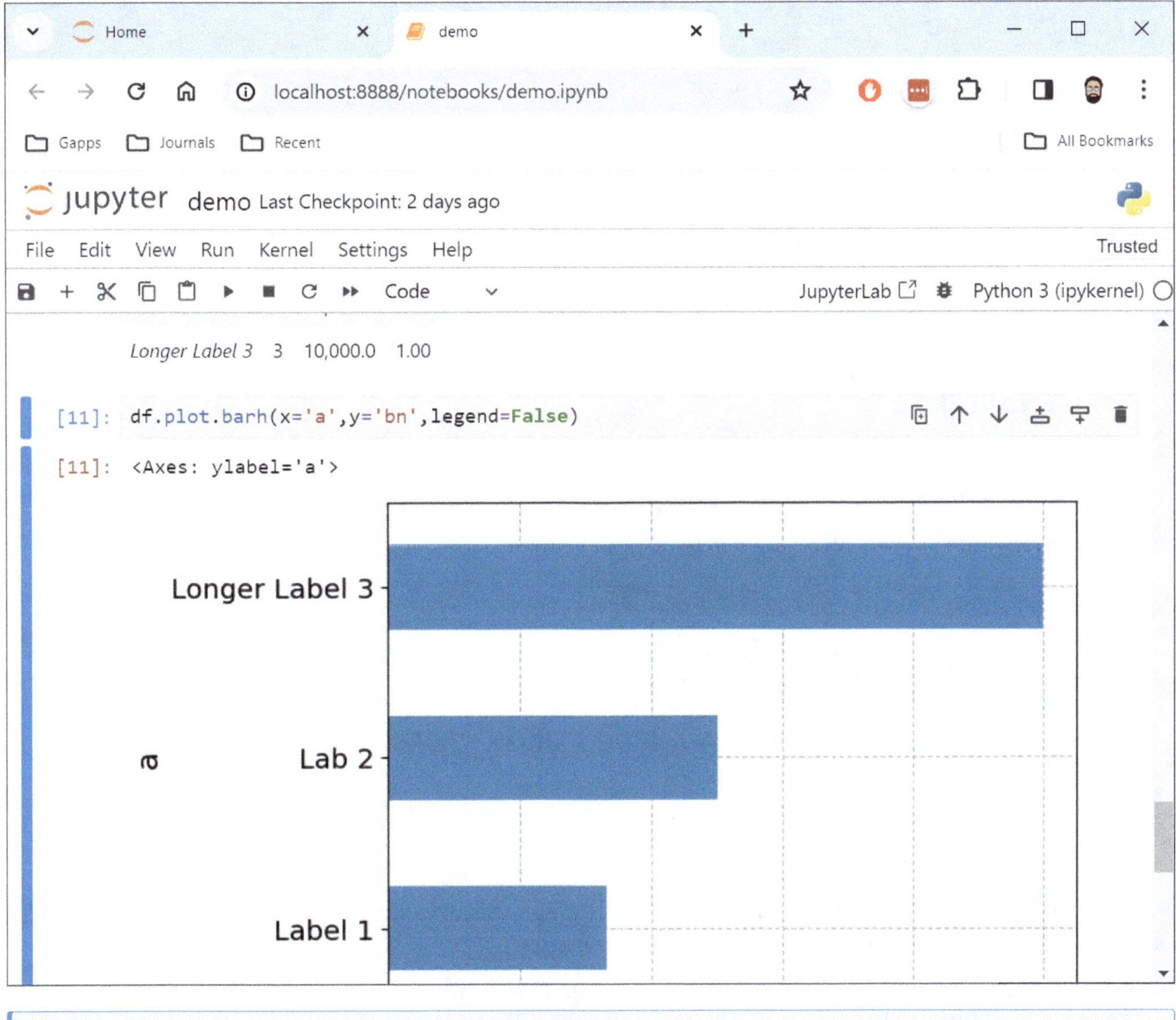

> Note
>
> Since I had you call `from src import mpl_looks` earlier in the script, this set my default chart template. If you did not call that, your chart may look different!

If you want to use the matplotlib figure/axes objects directly though, you can do that as well. Here is a smaller and more stretched out bar chart.

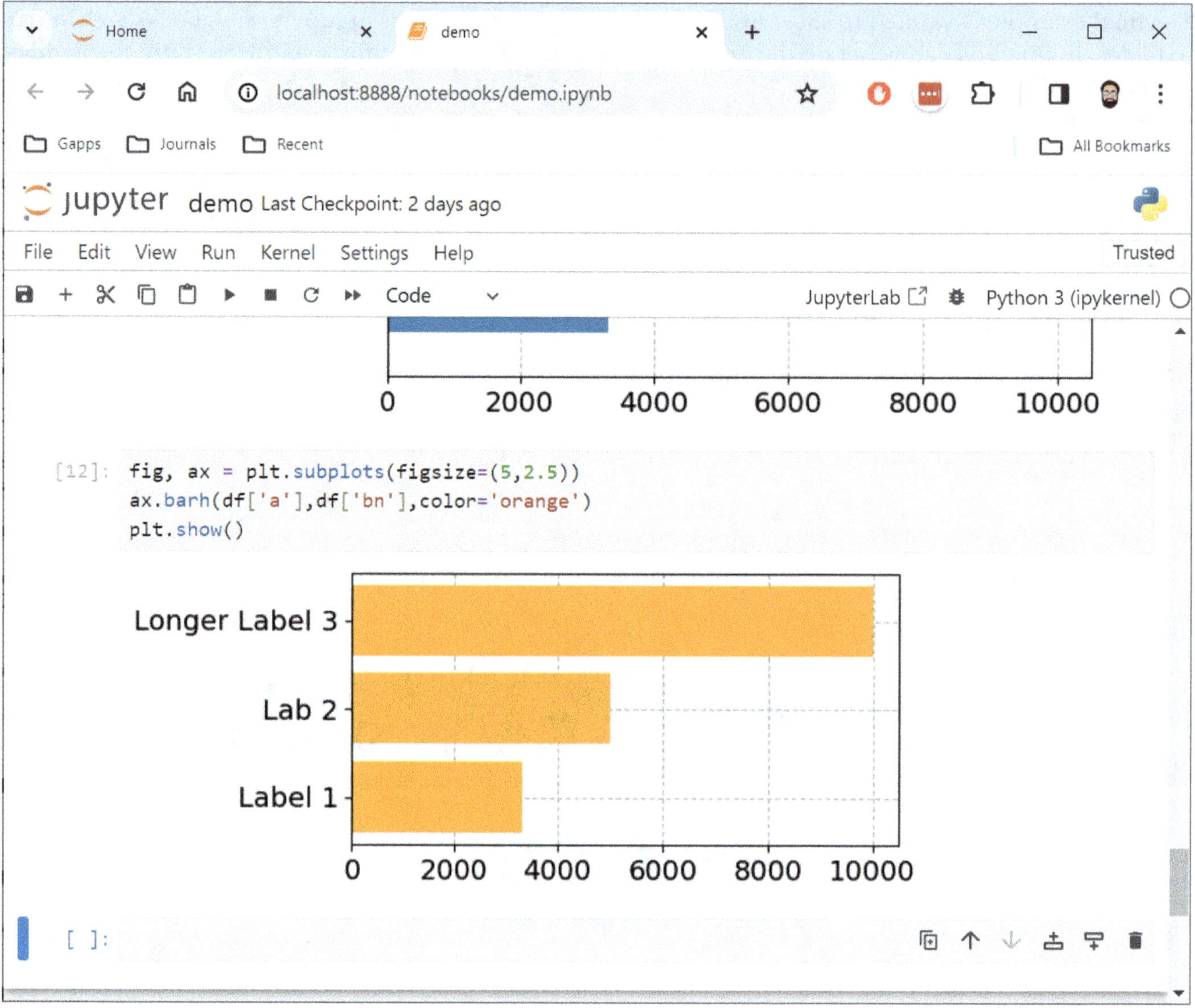

Here I use `plt.show()` instead of `fig.show()`, although this does not make a difference for any of the examples I show in either this chapter or the prior chapter.

You can also output multiple graphs in the same cell. This is not quite as flexible as the prior chapter where you can align graphs both vertically and horizontally in a specific arrangement, but looks more like you are just printing subsequent graphs one after another.

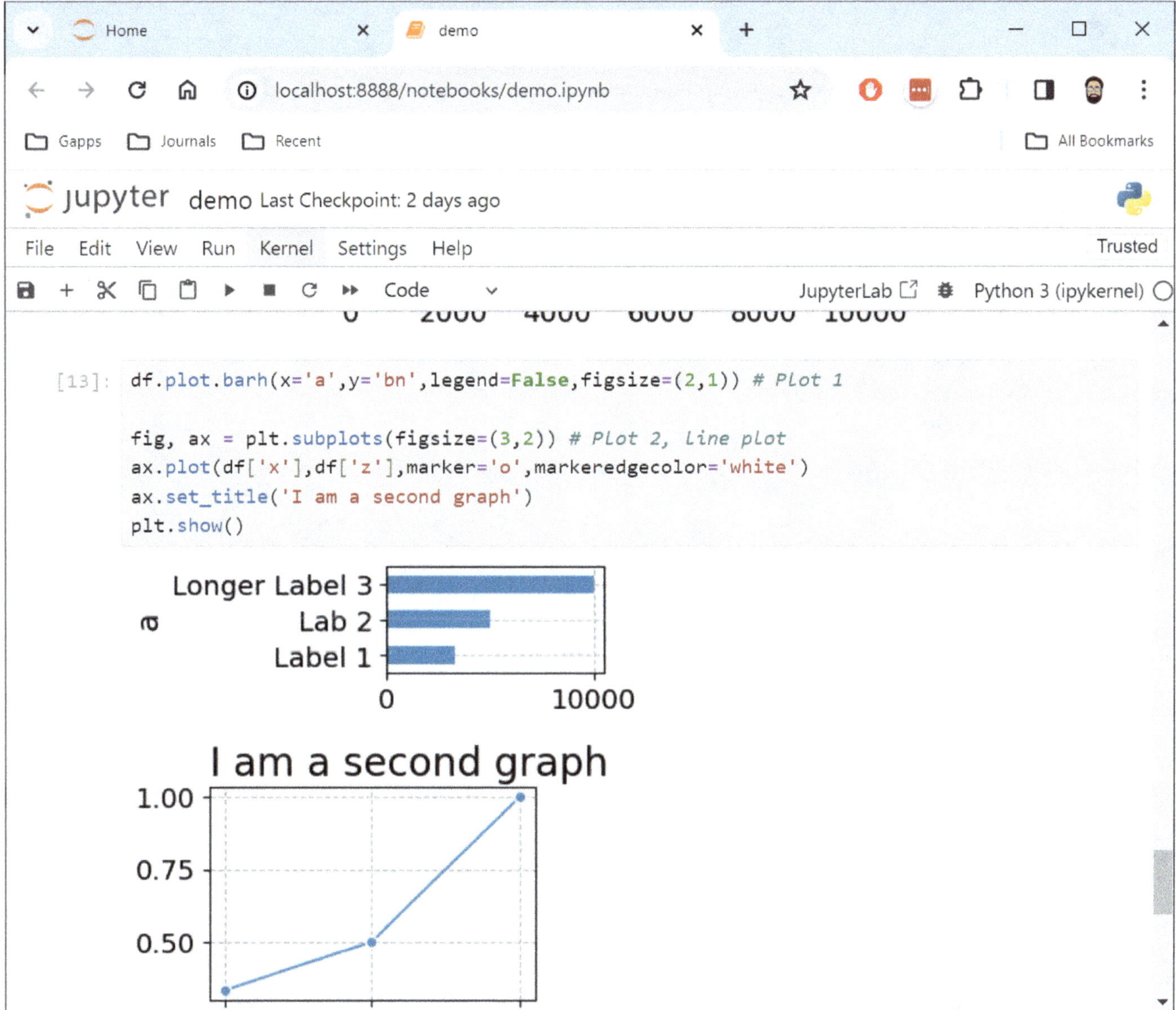

9.4 Executing Notebooks

So far I have shown executing individual cells one at a time. This is equivalent to going into a REPL session and conducting ad-hoc analyses. But, the main advantage of Jupyter over a REPL script is to intermix documentation. This allows you to create a finalized report that can be automated. Go ahead and save your notebook so far you have been running, then close out the file. Then back in the command prompt hit the keys `Ctrl` + `C` to stop the Jupyter server. Now from the command line execute the command `jupyter nbconvert demo.ipynb --to html`.

Once that is executed, in your folder with your notebook, you will then have a HTML file that is the contents of the notebook.

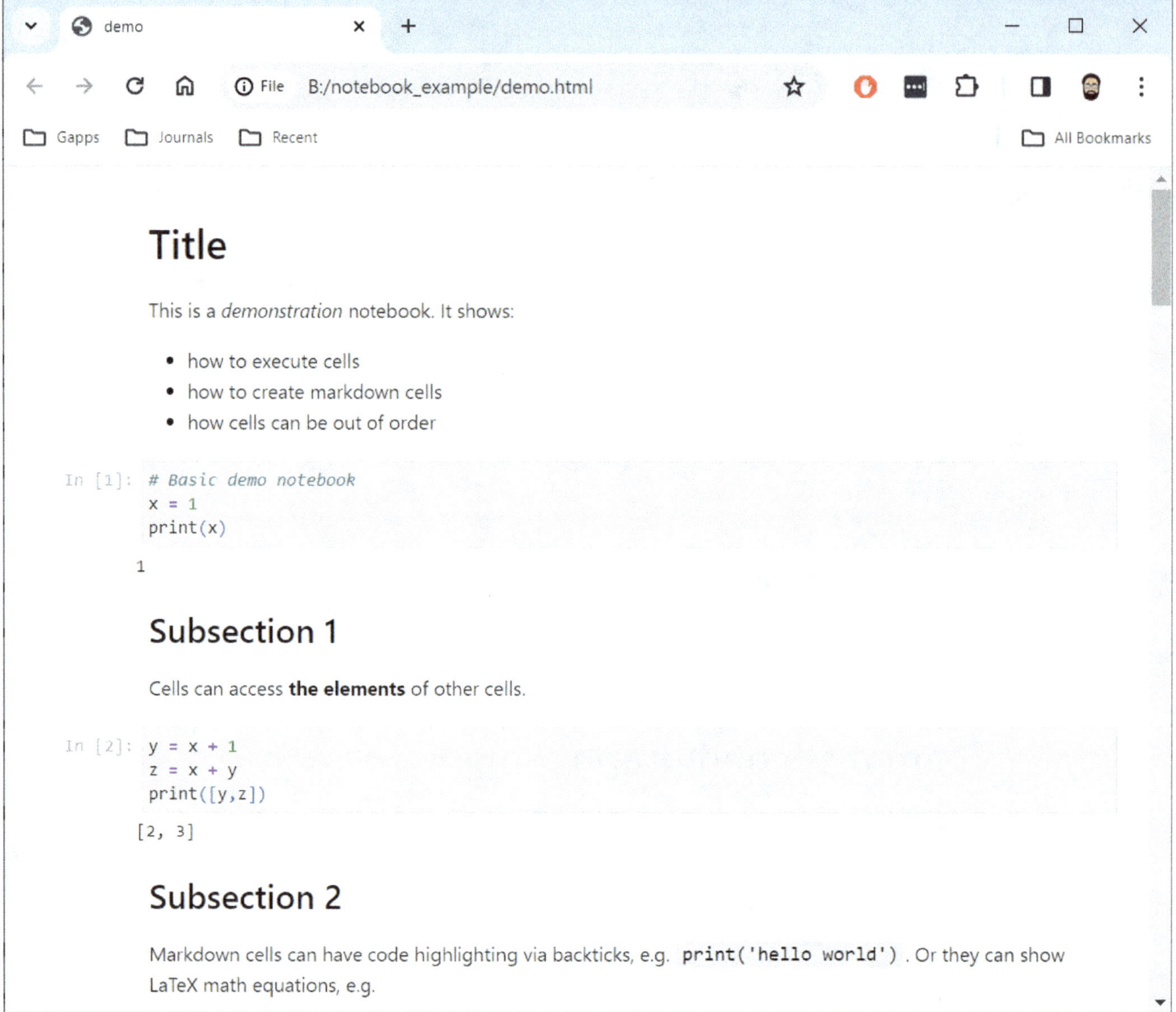

> Note
>
> You may need to install an additional library to have the `nbconvert` command line tool work, `jupyter_contrib_nbextensions`.

You can technically share this file, but it is similar to sharing an Excel spreadsheet – it may make sense to share this with another analyst, but to disseminate a general product this is not a good idea. First, you can remove the code from the final product by specifying the `--no-input` argument.

```
jupyter nbconvert --execute demo.ipynb --no-input --to html
```

The `execute` flag also re-calculates all of the cells in the notebook. So if you did not do this, it would simply output the current saved values in the notebook, and not re-calculate the cells.

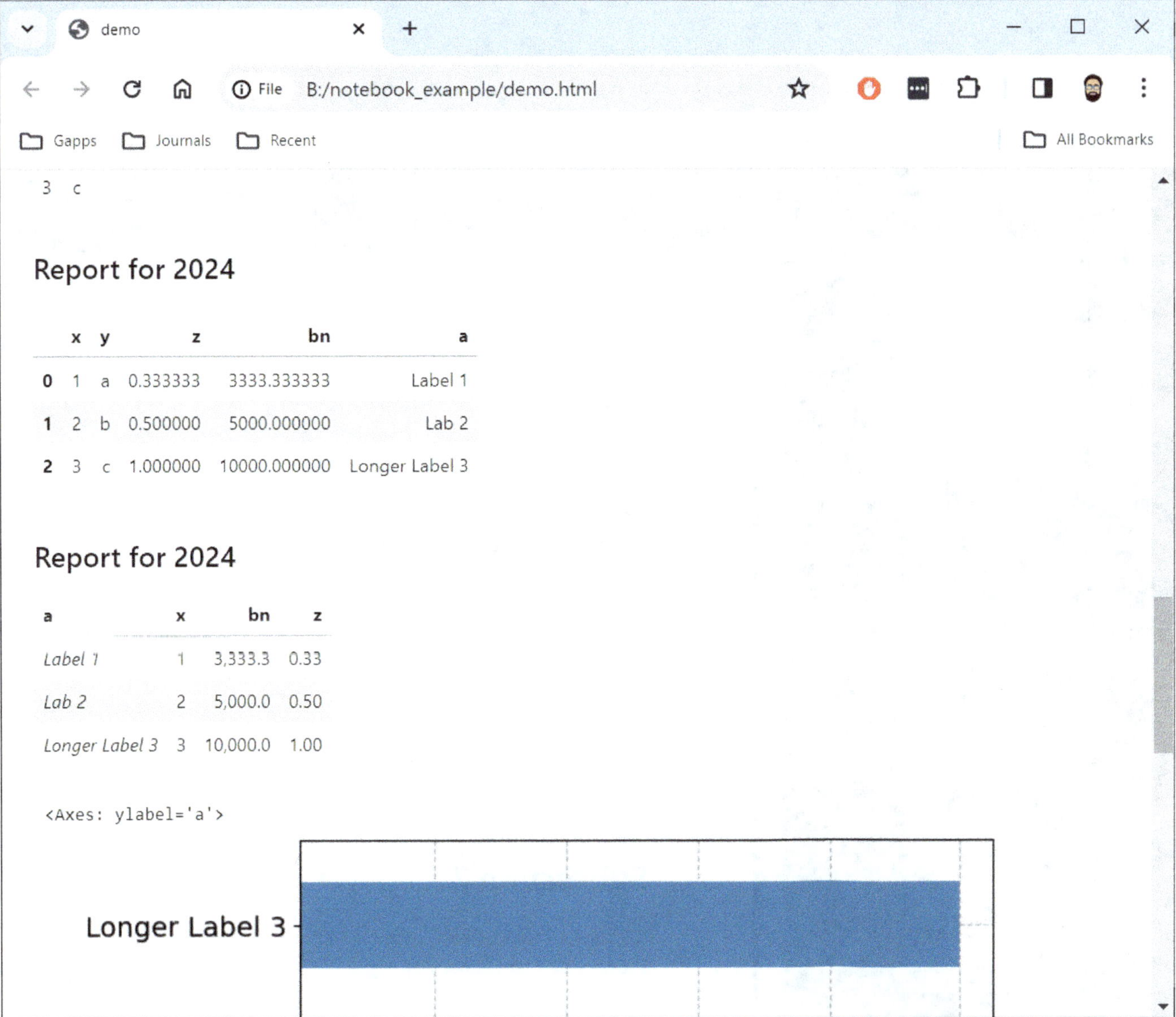

You can save the results to other formats, directly converting to PDF is perhaps the most commonly used to make reports more easily shareable.

```
jupyter nbconvert --execute demo.ipynb --no-input --to webpdf
```

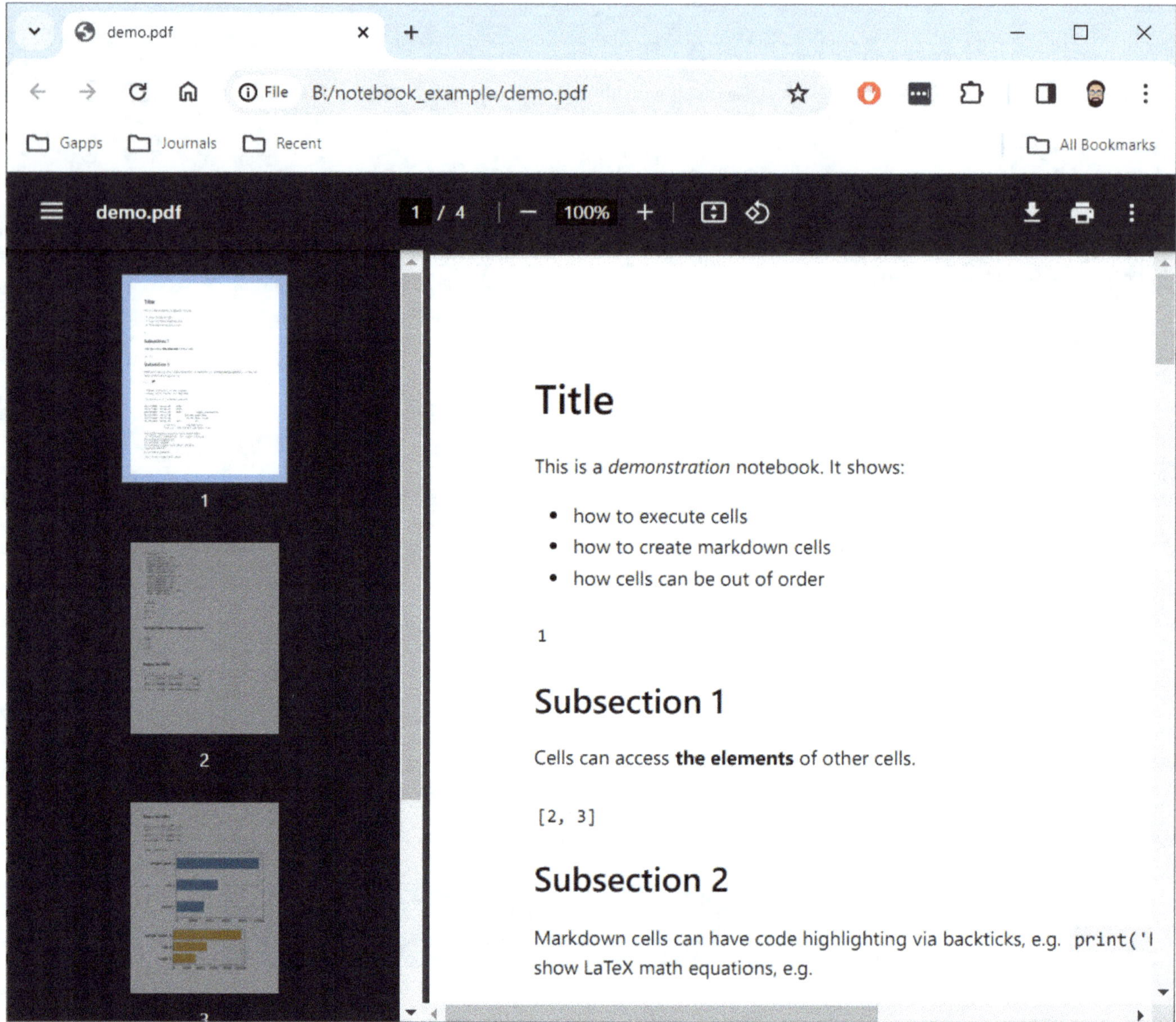

There is also an option `--to pdf`. This saves the notebook as a LaTeX file, and then converts the LaTeX file to PDF. The webpdf version is more akin to if you printed out a webpage directly. There are other output formats, such as reveal.js slides, or convert to markdown (which you then may use a different tool called *Pandoc* to convert to other formats, like epubs, Word documents, or PowerPoint presentations).

> Note
>
> Jupyter notebooks have undergone *many* changes, and the above nbconvert command may not work with your current version. My suggestion is if you need this functionality, is to experiment with creating new python environments to export notebooks. This example is done with `nbconvert` version 7.2.9 and `notebook` version 6.5.2. To get the `--to webpdf` to work, you may also need to install `pip install nbconvert[webpdf]`.

My experience is you can get reports like this to generate many of the elements you want, but depending on the level of control you want on the final product may dictate the overall workflow you use. For example, if you wanted a very nicely produced end of year report, you may have a notebook generate 50% of the report, but include that in an additional Publisher or InDesign document.

Note

An alternative to Jupyter notebooks is a program called Quarto, see https://quarto.org/. Quarto documents you can compile to Word documents or PowerPoint directly, which are easier to edit than other formats. (This book is created via Quarto.)

10 An End to End Project Example

The prior chapters have discussed individual elements of writing python code. The hardest part of using python in practice is not writing the individual code elements, it is creating them in a way that the pieces can function all together to achieve a particular end goal. This chapter is oriented around creating a simple *end-to-end* project that illustrates how different elements of python code co-exist.

The example project, which you can view and download online at https://github.com/apwheele/CrimeBook, is creating an automated CompStat like report, using crime data reported by the Cary North Carolina police department. The key elements of the project are:

- a README that describes the overall context of the work and key contact information
- a `requirements.txt` file that shows how to create the python environment necessary to run the code
- code (in the src folder) that has several functions in two different `*.py` files
- A Jupyter notebook, `CaryReport.ipynb`, in the root of the project, that generates the report

> Note
>
> if you are familiar with git and github, you can copy the project locally by running
>
> ```
> git clone https://github.com/apwheele/CrimeBook
> ```
>
> Or if you go to that url, you can download the files to replicate this chapter directly.
>
> Git is version control, and I think it is a very useful skill for analysts to be familiar with. I have notes for crime analysts getting started with git version control and github at https://andrewpwheeler.com/2024/01/14/getting-started-with-github-notes/.

This chapter will walk through all of the different key elements of the project, why they are structured like that, and how in the end you can create an automated report.

10.1 Project Setup

Before writing code, I wanted to go over the typical way a python project looks. This project, broken down to its core components, looks like below:

```
/CrimeBook
      |    README.md
      |    requirements.txt
      |    CaryReport.ipynb
      |
      ----/src
          |    prep.py
          |    viz.py
```

This tree diagram shows in the *root* of the project three files; a readme document, a requirements.txt file, and a Jupyter notebook.

> Note
>
> By root of the project, I mean its base folder. So on my Windows machine, I have this project currently in the folder `G:\CrimeBook`, and the readme document is then at `G:\CrimeBook\README.md`. It is common convention to expect scripts to be run so that the root directory of the project is the current directory of scripts.

The md file extension for the readme file is a *markdown* file, but that does not matter, it could be `README.txt` or `readme.txt` for example. The readme document should contain key information on the project. For crime analysts it may be the overall goal of the project, the person who made a specific analytic request, contact information for key individuals – anything you think could be helpful for either yourself or other fellow analysts in the future to understand the project. For python specific projects, it is common to have information here on how to create the appropriate python environment, and if any particular environment variables are needed (such as secrets to query databases, see Chapter 7 for discussion of secrets).

The readme document is metadata. It is not code, but useful information to understand the code more broadly and how to run the code. Similar to how you can write docstrings (information on your python functions), you should have notes on the overall project. Imagine if you had an accident tomorrow and a new analyst needed to take over your work – you should (to the best of your ability) have necessary metadata for the new analyst to replicate your work.

> Note
>
> There are more objects in the folder on github, including a license file and a `.gitignore` file. These are idiosyncratic to github and open source projects more generally. The license I use (MIT Open Source license) means you can copy the code if you want for use in your own projects, you do not need to obtain permission to re-use that code.

The second file I mention here, `requirements.txt`, includes the necessary python libraries that are needed to be able to run the code in the project. In Chapter 5 I discussed how to set up a local environment using the conda package manager. Here the `requirements.txt` file even has comments on how I use conda to create the environment.

```
# EXAMPLE COMMAND CREATING ENVIRONMENT
# conda create --name pybook python=3.10 pip
# conda activate pybook
# pip install -r requirements.txt

requests
pandas
tabulate
matplotlib
notebook==6.4.12
nbconvert[webpdf]
jupyter_contrib_nbextensions
```

The package relies on several libraries, with very typical ones used in data analysis, such as pandas and matplotlib. It also uses a slightly older version of Jupyter notebooks that is consistent with the nbconvert library (what we will later use to auto-convert the Jupyter notebook to a PDF document).

The last file in the root of the directory, `CaryReport.ipynb`, is the main executable intended to generate the final report. I will devote a section to more specifically detailing this file. But projects often have a *single* main function in the root of the directory. For scripts that are batch jobs, e.g. run this once a week, it is common to call that `main.py`, and so you schedule the command `python main.py` to run on a regular cadence. Another popular format is for live apps, like Flask apps, to have a file `app.py` in the root of the directory.

The way I have this project set up is not the only way to set up a project. For example you may have a folder for presentations or generated graph images, multiple notebooks in the root directory or in a different notebooks folder, other methods to create the environment (e.g. a python .cfg file or a docker file to set up a container), etc. But, this basic set up I have here I find very useful to start from, and then I expand out and change as needed for different projects.

10.2 Prepping the data functions

The folder src has two different .py files; prep.py and viz.py. These two files contain different functions, prep.py has functions to *prepare* the data (read the data from a web-api, filter for crimes of interest, and generate different tables). The file viz.py has functions that generate tables and graphs, different types of visualizations. You could technically have all of the functions in a single .py file, or each have each function have its own .py file. I often go with a middle approach, where I try to create functions in logical groupings. I will provide a brief rundown of the functions in each in turn.

For `prep.py`, it first loads the libraries I need, two new libraries I have not illustrated so far are the `StringIO` library and the `requests` library – these are both used to do a query of Cary's online API that has crime data. In most crime analysis applications, you would instead be reading from a local datasource, in which you might use the sqlalchemy library or the pyodbc library (see Chapter 7 for examples). But for this demonstration I am using up to date public data I can download from the internet to illustrate.

```
# This file preps the Cary data

from datetime import datetime
from io import StringIO
import numpy as np
import pandas as pd
import requests
```

The next part of the file has several objects (that are used in subsequent functions). It is common to put different fixed constants like this at the beginning of scripts. For example one of the dictionaries is a set of crime harm weights I used to rank order apartment complexes with the most problems later on. That dictionary is below, showing that I weight violent crimes as more serious than property crimes:

```
weights = {'Murder': (100,'Mur'),
           'Robbery': (25,'Rob'),
           'Agg Assault': (15,'AA'),
           'Burglary': (12, 'Bur'),
           'MV Theft': (10, 'MVT'),
           'Theft from MV': (5, 'TFMV')
}
```

> Note
>
> There are many different crime weighting schemes, but they are often highly correlated with one another. See for example my blog post https://andrewpwheeler.com/2020/05/22/conjoint-analysis-of-crime-rankings/ illustrating two different harm weighting (one from survey data, another set of scores developed by Jerry Ratcliffe). You can often make ad-hoc choices, and the different rankings will have similar results in places or people that rank as having the highest harm scores.

The reason I list this as a constant at the beginning of the file is that it is likely to be used in multiple places in future additions of the code base. So this code has an example of aggregating up crime harm weights to apartment complexes, but in the future you may wish to do this for other addresses, or for individuals (as is common for chronic offender lists). Having this as a constant at the top of the file allows future updates to easily edit this constant, e.g. if you want to add in gun crimes to the harm weights, or if you want to change the weight values, e.g. change theft from MV to be equal weight to burglary.

If you edit the `weights` object once, it will then propagate to all the other functions that use this information. So this is easier in lieu of of defining the weights multiple times in code, and when updating having to edit multiple functions (or worse, forgetting to edit and having varying definitions).

Then it begins the functions, the first function is `get_cary`, which loads the Cary crime data:

```
def get_cary():
    '''function to get Cary online data'''
    cary_url = ('https://data.townofcary.org/api/explore/v2.1'
                '/catalog/datasets/cpd-incidents/exports/csv?'
```

```
                 'lang=en&timezone=US%2FEastern&use_labels=true'
                 '&delimiter=%2C')
    res_csv = requests.get(cary_url)
    data = pd.read_csv(StringIO(res_csv.text),low_memory=False)
    return data
```

Again, for most analysis it will be `pd.read_sql` instead of `pd.read_csv` (and pulling data from a webapi). But the logic may be the same.

> Note
>
> Depending on the nature of the analysis, you want to prevent generating big queries repeatedly. One approach to this may look like:
>
> ```
> # Example of caching data
> from datetime import datetime
> import os
> import pandas as pd
>
> file = "./src/cpd-incidents.csv"
> ft = datetime.fromtimestamp(os.path.getctime(file))
> nw = datetime.now()
> elapsed = (nw - ft).total_seconds()
>
> # if older than 1 day, update the data:
> if elapsed > 60*60*24:
>     data = get_cary()
>     data.to_csv(file)
> else:
>     data = pd.read_csv(file)
> ```
>
> What this code does is looks at the file `cpd-incidents.csv`, and if the file is more than one day old, it redownloads the data. Else it just reads in the local data file. This is very convenient for debugging code, when the initial query takes a long time to run.

The second function, `prep_data`, calls the first function, and then does what is commonly called feature engineering. It takes in the crime data, identifies a few different key categories of interest, and turns them into dummy variables (so in the `Robbery` column, if a row is a robbery it equals 1, and 0 otherwise). This will be useful for later functions that aggregate crimes to different groupings. The function also calculates the most recent data in the data – this will be useful later on to make notes in our tables and graphs for when the code was run. Say you generate a PDF report and email it to the command staff, it will be useful in the report to have when the numbers were calculated, as they will ultimately change over time (either based on more recent data, or historical crimes may be retrospectively edited).

I have edited out some of the docstrings to make the function shorter, and do not describe fully the process of creating the dummy variables. It uses loops over the dictionary `cats`, which is defined at the beginning of the file, as well as other functions shown in different chapters, such as if-else statements (Chapter 2 if statements, Chapter 6 on pandas dataframe manipulation).

```
def prep_data(cat_info=cats,
              extra_vars=['Apartment Complex',
                          'Phx Community',
                          'Phx Status']):
    # This downloads data and creates dummy variables
    data = get_cary()
    for cn in cat_info.keys():
        c1, c2 = cat_info[cn]
        if c2:
            data[cn] = ((data['Crime Category'] == c1) &
                        (data['Crime Type'].isin(c2)))
        else:
            data[cn] = (data['Crime Category'] == c1)
    # converting to timevariable
    data['Date'] = pd.to_datetime(data['Begin Date Of Occurrence'])
    # getting last date in data
    max_date = data['Date'].max()
    # Returning only variables I want and rows
    # that meet one of these criteria
    dummy_vars = list(cat_info.keys())
    any_vals = (data[dummy_vars].sum(axis=1) > 0)
    data[dummy_vars] = 1*data[dummy_vars]
    keep_vars = ['Date'] + dummy_vars + extra_vars
    # sorting so oldest to newest
    data.sort_values(by='Date',inplace=True)
    data = data.loc[any_vals,keep_vars].reset_index(drop=True)
    return data, max_date
```

The function has a set of defaults that creates many regular crime type dummies. But imagine in the future you want to expand the analysis to specifically look at commercial burglaries with latitude and longitude data to create maps, you may then pass in the correct dictionary format to the `cat_info` argument to only pull out commercial burglaries, as well as add in the latitude and longitude variables to the `extra_vars` argument. These functions are being created now to help make a single report, but it is a good idea to make those functions more flexible when it is reasonable to do so.

Now when one runs this function, it will return a dataframe that looks like:

Date	Murder	Agg Assault	Burglary	...	Apartment Complex
1989-08-05	0	0	1	...	nan
1994-07-10	0	1	0	...	OXFORD APTS.
1999-10-06	0	0	1	...	nan
1999-12-10	0	0	1	...	nan

Where we have a date variable, several dummy variables for different crime types, and then a few additional variables related to apartment complexes. Based on this pre-processed data, one then has several functions to return different aggregations. There are three additional functions in prep.py; `ytd_stats`, `month_counts`, and `apt_metrics`. I will not review them line by line here (I encourage the interested reader to go and see how I created these), but again they are simply chaining together different operations I have shown in prior chapters to calculate fairly regular metrics crime analysts are interested in.

> Note
>
> Cary's online crime data is supposed to be back to the year 2000. As you can see in this sample of data the date field, which is derived from the begin date of the crime occurrence, has some outlier values much earlier. This can happen if one is using different dates for selection, such as the report date in one instance and the date the crime is thought to have occurred in another, that is someone reports a crime far after it has occurred.
>
> For different products the analyst may wish to use one or the other. A benefit of code is that you can see how particular metrics were calculated, so if someone else says "there are 2 murders so far this year" and your code says there are 3, you may be able to identify the root cause of the difference due to using such different definitions.

Once you have this set of data, I then have three additional functions in `prep.py`. Those functions are `ytd_stats`, `month_counts`, and `apt_metrics`. I do not copy them here (but encourage the interested reader to go look at them and see how they operate). They calculate in turn year to date stats (comparing the current year to the prior year), monthly counts (for the prior three years), and aggregate crime harm weights in the past 90 days for apartment complexes.

These are regular types of analyses that are typical of CompStat like reporting.

10.3 Creating the visualization functions

The second script in the src folder, `viz.py`, first sets the matplotlib theme, see Chapter 8 for discussion. Second it has four functions, one for each subsequent cell in the Jupyter notebook for the final report. For example, the first function creates the introduction to the report. One could place a markdown cell in the report directly, but that does not allow for dynamic information, like the dates the report was run. So the function `intro_slide` allows one to dynamically insert that information, along with a contact.

```
# Default contact info
contact = ('Andrew Wheeler', 'andrew.wheeler@crimede-coder.com')

# Intro title slide
def intro_slide(date,cont=contact):
    intro  =  f'# Cary Monthly Crime Report\n\n'
    intro += ("This is an automated CompStat report that provides "
              "up to date metrics for ")
```

```
    intro += "\n - the most recent year-to-date metrics"
    intro += "\n - monthly graphs over the prior three years"
    intro += ("\n - The top 10 apartment complexes with the highest "
              "weighted crime harm scores over the past 90 days")
    intro += f"\n\nLast run at {date.strftime('%Y-%m-%d')}\n"
    intro += f'By [{contact[0]}](mailto:{contact[1]})'
    return intro
```

Two things for this function, for most analysts you will be writing code you interact with yourself, so it may make sense to hard code the contact information. If you are in a crime department where you are the sole analyst, the contact will always be yourself. But, since this may need to be changed in the future, I think it makes sense to place the main contact as a constant at the top of the file (similar to the harm weights in the `prep.py` file). It is also the case that if you have a set of *shared* code among all crime analysts, this is useful as it allows it to be changed from report to report. Having a parameter in the function to alter this if necessary, but having a default value, is a simple approach to get both ease of use (for the single analyst that uses it most frequently) as well as potential flexibility if needed.

The other main set of text that this function inserts is a passed in date. You could do the date inline, for example have `datetime.now().strftime('%Y-%m-%d')` inside of the function, so it dynamically inserts the date at run time. I did not do that here, as the Cary data can sometimes be lagged, so I want the date to reflect the last date in the reported crime data, not per se the date the report was run.

One could parameterize this even further, such as having the ability to pass in a title or other additional text. If you were re-using this library to conduct various ad-hoc analyses, not just a single CompStat report, that would be a useful update to this function.

The next function, `ytd_ouput`, is quite simple. It takes in a dataframe, a date string, and returns a set of concatenated markdown text. See Chapter 9 on formatting markdown tables in Jupyter notebooks. This is just a helper function to do it all at once, in the particular format I want for my year-to-date table in the report. The function returns a string, which I will then later render in the Jupyter notebook using the `Markdown` function.

```
# Year to Date Table
def ytd_output(data,date_str):
    title = f'## Year to Date Stats as of {date_str}'
    mt = data.to_markdown()
    return title + '\n' + mt
```

The next function, `month_graphs`, generates a set of multiple plots, looping over the different columns in the passed in dataframe (assuming the first column is the month, and subsequent are the counts). First, I use a trick to only show the month labels for January and July, since the charts are in a smaller space. I also then rotate those months and use a slightly smaller font. I intentionally set the size of the figures so I could fit two figures per page. But this function will produce how ever many graphs you feed it from the dataframe, it is not a fixed number of graphs. This is something that is very difficult to do in other report creating applications, do loops over data. But this example is fairly simple to accomplish.

```
# Monthly Graph last 3 years
def month_graphs(data):
    cols = list(data)[1:]
    # set xticks for January and July
    tr = data['Date'][data['Date'].dt.month.isin([1,7])]
    tr_labs = tr.dt.strftime('%Y-%m')
    for c in cols:
        fig, ax = plt.subplots(figsize=(4,1.9))
        ax.plot(data['Date'],data[c],'-o',c='k',markeredgecolor='white')
        ax.set_xticks(tr,tr_labs,rotation=30,fontsize=10)
        ax.set_title(f'{c} counts per month')
        plt.show()
```

When running this code inside of a notebook, doing `plt.show()` will insert the graph into the notebook cell. If intending to run similar code in non-notebook environments, you may wish to have that in an additional if block, e.g. have a parameter `show=True`, and then in the code `if show: ... plt.show()`. You can of course have more complicated logic as well, such as saving the individual graphs to different PNG files (to insert into other reports or formats if you like).

The last function, `apt_table`, is similar to the year to date function, except that it does two different data manipulations to the input dataframe. One is that it takes the top n number of apartments. If I included the full table in the report, it would likely stretch out over several pages, and include many apartment complexes with no crime. Choosing the Top 10 is a common analytical tool to focus on the places that have the most problems, while still making the report readable. I set n as a parameter in this function, incase in the future I wish to include the top 15 or top 20 or top 5 – having a parameter makes this easily modifiable.

A second data manipulation I conduct is to *escape* the pipe character. The pipe character happens to be a special character in markdown, so the note cell (that concatenates the different crime subsets into a smaller note that looks like `Rob 1 | TFMV 1`, will turn into `Rob 1 \| TFMV 1` after the data manipulation.

```
# Apt Crime Weights Table
def apt_table(data,range_str,n=10):
    topk = data.head(n).copy()
    # escaping the pipe in the markdown for the note
    topk['Note'] = topk['Note'].str.replace(r"|",r"\|")
    fin_str = f'## Top {n} Apartments by Crime Weights\n\n'
    fin_str += range_str
    tkm = topk.to_markdown(index=False)
    fin_str += "\n\n" + tkm
    return fin_str
```

10.4 The Jupyter notebook automated report

The final part of this example project is the code that creates the report, the `CaryReport.ipynb` file. This notebook only has four cells, and is displayed in the screenshot below.

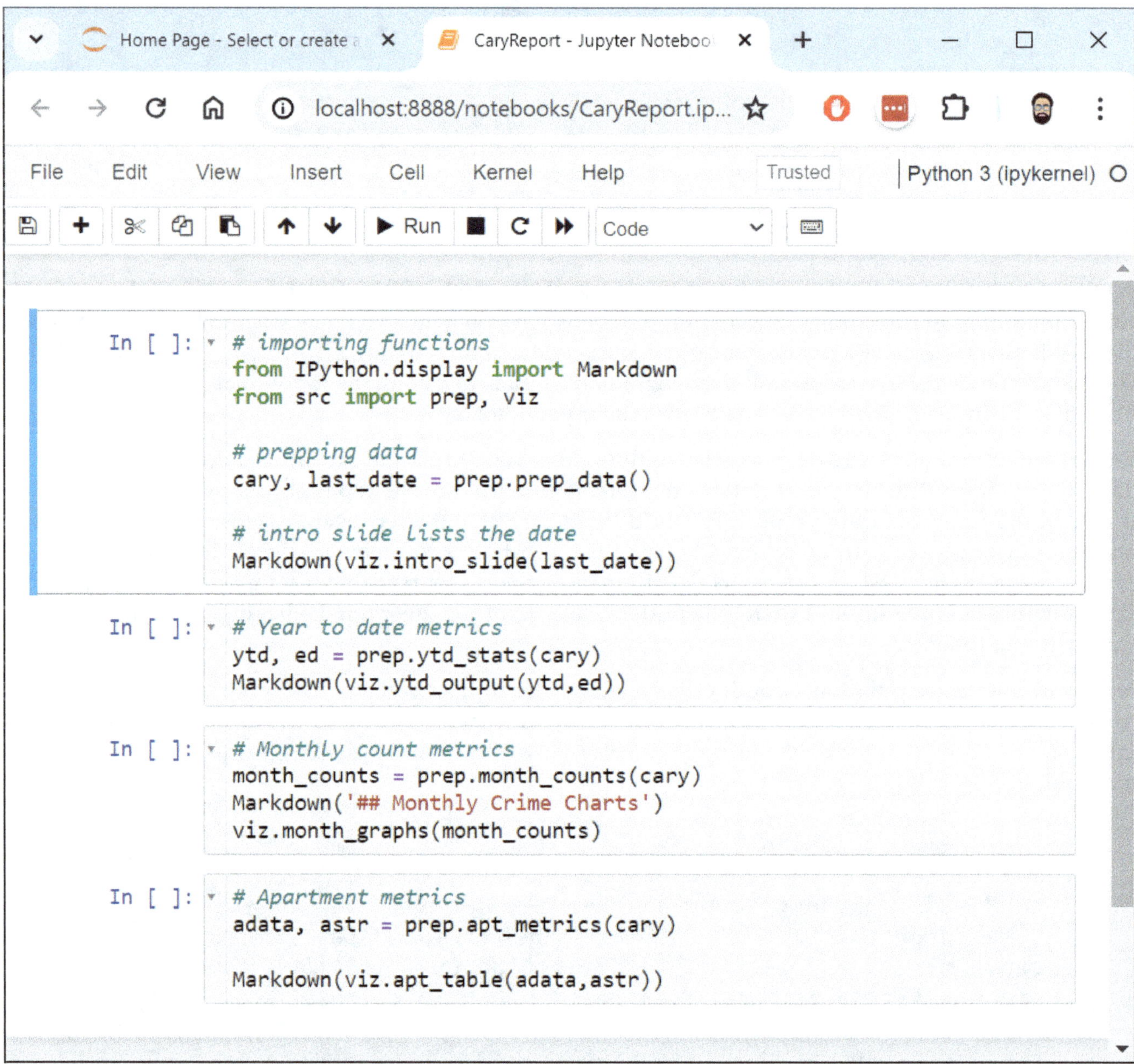

The first cell imports the `Markdown` function, and imports our local functions `prep` and `viz` we just discussed (that are in our `src` folder). Then we get the updated Cary data, including the last date in the data, and then create our intro slide on the final line.

Then each cell is subsequently one of the prep + viz functions to produce a final output. The second cell is year to date metrics, calculating the stats from the `cary` dataframe, and visualizing the markdown results. The third cell is the monthly crime counts, which I include markdown text directly to title that section, but then call the `viz.month_graphs` function to generate the charts in a loop. The fourth and final cell is the apartment metrics.

Most analysts will want to create a PDF report that can be printed and emailed, to do that with the Jupyter notebook, one can then run the command:

```
jupyter nbconvert --execute --no-input --to webpdf CaryReport.ipynb
```

> Note
>
> You may also need to include the flag `--allow-chromium-download` at least one time when running the code. This code renders the HTML in a web browser and prints to PDF from the web-browser. Chromium is a tool to do this using Google chrome's web-browser, and may need to be downloaded on the local system.

You can see an example of the final version of this PDF at https://github.com/apwheele/CrimeBook/blob/main/CaryReport.pdf. I have placed those notes in the README document, to again help future analysts if they wish to generate a similar report. Here is a screenshot of the first page of the report:

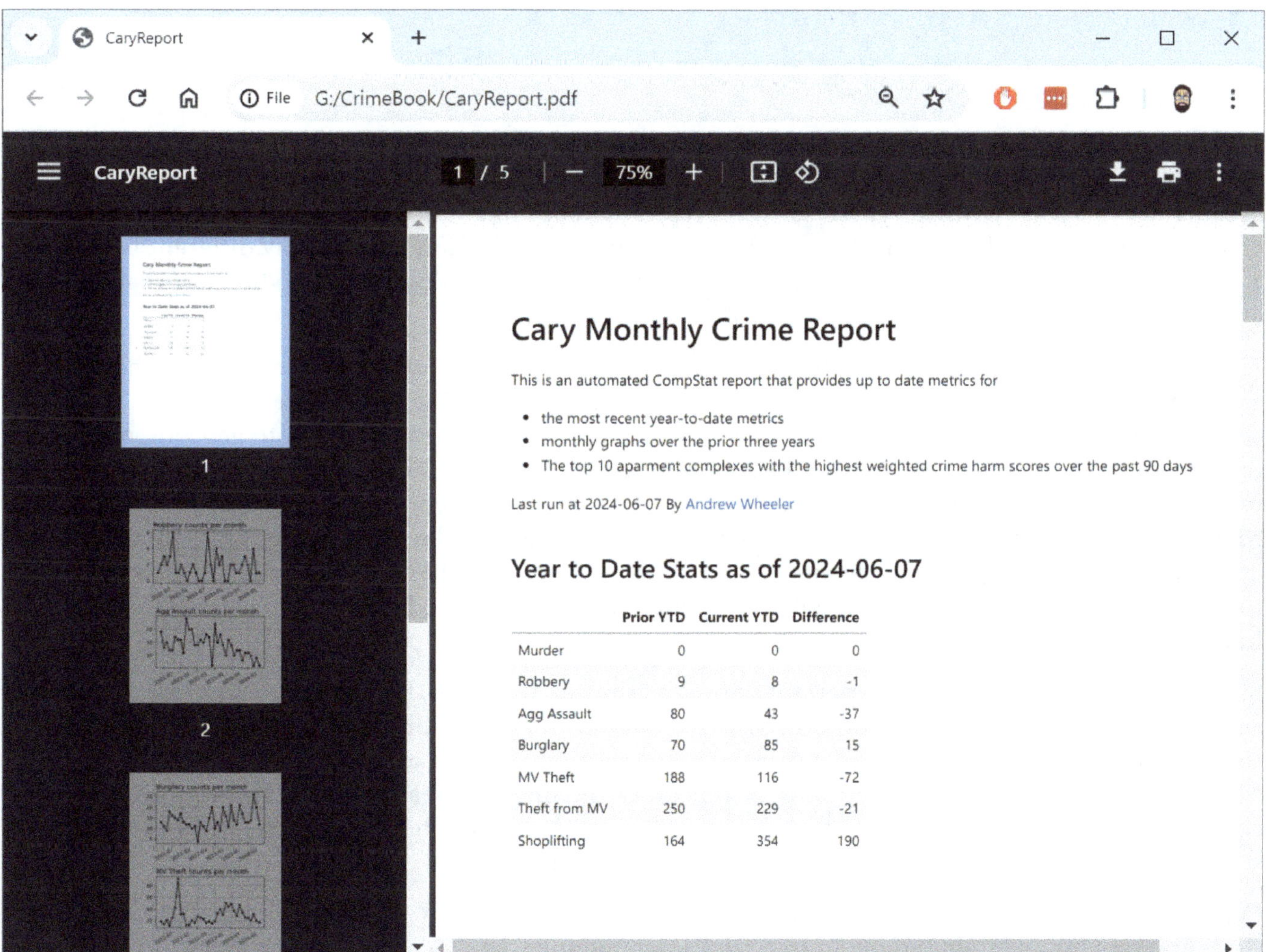

Cary Monthly Crime Report

This is an automated CompStat report that provides up to date metrics for

- the most recent year-to-date metrics
- monthly graphs over the prior three years
- The top 10 aparment complexes with the highest weighted crime harm scores over the past 90 days

Last run at 2024-06-07 By Andrew Wheeler

Year to Date Stats as of 2024-06-07

	Prior YTD	Current YTD	Difference
Murder	0	0	0
Robbery	9	8	-1
Agg Assault	80	43	-37
Burglary	70	85	15
MV Theft	188	116	-72
Theft from MV	250	229	-21
Shoplifting	164	354	190

If wanted, an analyst could then create a batch script to auto-create this report on a regular basis, such as on the first day of the month. And do more complicated things if wanted, such as automatically emailing out the report.

> Note
>
> I have notes on automating reports using Windows bat files at https://github.com/apwheele/Blog_Code/tree/master/Python/jupyter_reports. Once you have a bat file that executes the report, you can use the Windows task scheduler to run the report at a specific time. On Mac machines you will use the crontab tool to similar effect.

One last item I want to note is that many analysts create very detailed Jupyter notebooks, often dumping *all* of the code into the notebook, instead of having a more minimal report. The name Jupyter notebook came from mimicking a laboratory notebook. I personally do not think that is a good idea, and it is better software practice to isolate your functions in more modular functions. This makes it much easier to modify the report at a later date. If you for example had the `get_cary` function defined inside the notebook, but then had many reports using that same copy-pasted function, if you needed to modify `get_cary` you would need to modify the function in each separate report. Whereas having the function isolated, you only need to modify it one time and it will be modified for all future reports.

10.5 What next?

While this chapter shows a brief example of generating an automated report, I want analysts to get a broader taste of what writing python software typically looks like. Those skills will be useful to not only create reports, but to do various data analysis and data manipulation tasks. Many different tools could create a similar report to what I show here, the power in understanding python code is not that it can do one specific thing, but that *it can do many things*.

This chapter shows how to create a PDF report, but if you wanted to create an interactive dashboard you can do that (see my example Dallas Crime trends dashboard using the panel library at https://crimede-coder.com/graphs/Dallas_Dashboard). If you want to script an automated job to geocode new crime records and save them in a table you can do that (see my example of using the ArcGIS python libraries to do that at https://crimede-coder.com/blogposts/2024/LocalGeocoding). If you want to do more complicated analyses, like predictive machine learning models, social network graphs, or semantic search of documents using large language model embeddings, those are all examples a crime analyst may wish to create in python. If you can do it with a computer at all, it is very likely you can accomplish that task in python.

Since report writing is such an important part of a crime analysts job, I am going to describe a few more resources specific to report writing I believe many analysts would find value in learning. Although I showcase Jupyter notebooks in this book, most of my personal work uses a different tool to render reports – Quarto documents, see https://quarto.org/. Under the hood it uses Jupyter notebooks to render the contents, and this book is even a Quarto document. Understanding how notebooks work is still a good starting place, for Quarto or other notebook like environments.

Quarto is convenient for several reasons, it can output content in word documents or powerpoint presentations for example (which I often find convenient to do semi-automated reports, so the report is filled in with various graphs and tables, but I need to manually review and add in additional text or other diagrams). Quarto can also pass in additional parameters to a report – say I wanted to generate a crime report for each individual area of the city. Instead of doing a new Jupyter notebook for each area, I could have a parameter and then loop over those different areas of the city using Quarto, see https://quarto.org/docs/computations/parameters.html for examples.

Because this is an entry level book, there are ultimately many different aspects of more advanced python programming I have not covered in the book. Doing more advanced numerical analysis is one, such as predictive machine learning models. While often included in introductory data science books I have intentionally not covered this material. Most analysts need help with making a top 10 crime hotspot areas, and learning machine learning is overkill. (Even though I personally know how to do predictive models, I am not above suggesting you should use a top 10 ranked list in many situations!) If you would find value in such a book, do not hesitate to contact me and let me know, https://crimede-coder.com/contact. Feedback like that is important to me to be able to understand when there is a market for such material.

There are additional data topics that do not come up for basic crime analysis, but are good skills to know to level up your capabilities to a more advanced data science role. These include more advanced SQL and deploying databases, geospatial analytics, and web-based applications. Python is not just used for data analysis, but it is used in such a wide array of applications that there are often convenient libraries and frameworks to do any of this work. Because python is open source, you can use those libraries for the most part with zero budget, and can extend applications to your personal projects. To learn those more advanced materials, the developer should have a good grasp of the basics. And this book hopefully provided that base.